Instrumented Systems for Microbiological Analysis of Body Fluids

Authors

Lorraine S. Gall, Ph.D.
Associate Professor of Epidemiology
Department of Virology and Epidemiology
Baylor College of Medicine
Houston, Texas

William A. Curby, M.S.
Head, Sias Biophysics Research Unit
Lahey Clinic Foundation
Boston, Massachusetts

CRC Press
Taylor & Francis Group
Boca Raton London New York

CRC Press is an imprint of the
Taylor & Francis Group, an **informa** business

First published 1980 by CRC Press
Taylor & Francis Group
6000 Broken Sound Parkway NW, Suite 300
Boca Raton, FL 33487-2742

Reissued 2018 by CRC Press

Library of Congress Cataloging in Publication Data

Gall, Lorraine S
Instrumented systems for microbiological analysis of body fluids.

Includes bibliographies and index.
1. Medical microbiology--Instruments. 2. Body fluids--Microbiology. 3. Body fluids--Examination. 4. Medical microbiology--Automation. I. Curby, William A., joint author. II. Title.
QR67.G34 616'.01'028 79-706
ISBN 0-8493-5681-4

A Library of Congress record exists under LC control number: 79000706

ISBN 13: 978-1-315-89463-8 (hbk)
ISBN 13: 978-1-351-07373-8 (ebk)

Visit the Taylor & Francis Web site at http://www.taylorandfrancis.com and the
CRC Press Web site at http://www.crcpress.com

PREFACE

The need for improved microbiological techniques for analyzing body fluids, which combine speed and accuracy at a reasonable cost, is evident to serve as a guide to the doctor in treating his patients adequately and more promptly. The first step in speeding up microbial analysis was a series of kits, mainly intended for identifying enteric bacteria that did not completely meet the needs for improving microbial analysis. In about 1965, instrumented systems for microbiological analyses of varying degrees of sophistication began to appear, but these did not make a successful market penetration. Beginning in 1970 several sophisticated decision-making instrumented systems were displayed at scientific meetings and were the subject of numerous scientific papers. Six of these systems capable of detection, enumeration, antibiotic susceptibility testing or identification are on or approaching the market. As a result of these activities, microbiologists are faced with evaluating these instruments for possible use in their laboratories and bioengineers are increasingly being called upon to design these systems. To intelligently design or evaluate these instrumented systems, information on the theoretical and practical aspects of instrumented microbiological analytical systems is necessary. It is the purpose of this monograph to set forth the theoretical rules and practical considerations as a basis for decisions, and to illustrate the application of these aspects to the design and evaluation of instrumented microbiological analysis of body fluids.

The first two sections of the monograph discuss the theoretical rules and practical aspects involved in instrumentation of a microbiological assay of body fluids. Following a brief discussion of several early systems that did not receive market acceptance, there is an in depth illustration of the pulse height analyzer as an instrumented system designed especially to apply these theoretical and practical considerations. The next six chapters describe various instrumented systems illustrating several different approaches to microbiological analysis of body fluids, followed by an evaluation of how well these instrumented systems reach the goals for yielding ideal analytical results. The last two sections deal with some less sophisticated instruments, and with the possibilities for the future in microbiological assays using sophisticated decision-making instrumented systems.

Lorraine S. Gall
William A. Curby

THE AUTHORS

Lorraine Sibley Gall, Ph.D., was born in 1915 in Binghamton, New York, where she graduated from high school as valedictorian, and then enrolled in the bacteriology course at Cornell University, receiving a B.S. degree in 1938. For the following six years she conducted research in nutritional microbiology, and then returned to Cornell to study for a Ph.D. in that subject, submitting a thesis on rumen microbiology. In 1947 she continued her studies with a post-doctorate fellowship at Yale University.

The next ten years were spent in productive research on the isolation and basic function of anaerobic rumen bacteria, first at Ohio Agricultural Experiment Station, and continuing at National Dairy Research Laboratory on Long Island, where her fistulated cattle, Christopher and Isabella, were curiosities and inspired front page coverage of her work in the New York Herald Tribune. Her recognition as an authority on rumen bactera led to the award in 1954 of a Senior Research A Fulbright grant to New Zealand and Australia, where she studied the influence of pasture on rumen flora and lectured on her techniques for isolating rumen anaerobes.

Returning to the United States, she found her strong convictions, that high-roughage low-concentrate feeding produced the most economical, efficient rumen function were not welcomed or accepted by the grain-oriented economy of the U.S. Dr. Gall then turned her attention to human microbiology. Intrigued by the trip to New Zealand, she combined business with travel and trained detailmen throughout the world for American Cyanamid, Lederle International Division; but after five years living out of a suitcase and several unpleasant incidents in an increasingly hostile world, she returned to research.

During the next ten years, while working for Republic Aviation Corporation and IBM, she participated in research programs related to the man-in-space project, monitoring the effect of various simulated space environments on the bacteria living in or on the body of man, with emphasis on the influence of diet on the intestinal anaerobes. Manually monitoring the microbiological status of the space capsule during flight is complicated by the difficulty of performing microbiological procedures in a weightless environment. This led Dr. Gall to investigate the automation of the microbiological analyses, and this has been her chief interest for the past ten years.

Her present assignment at Ames Division of Miles Laboratories is to continue the work started at Grumman Health Systems on an automated microbiological instrument that will perform a complete analysis in a matter of hours rather than days.

At present she is working on development of biomedical instrumentation as an Associate Professor of Epidemiology at Baylor College of Medicine in Houston, Texas.

Dr. Gall has published about 100 scientific papers and many popular articles. She has been recognized by inclusion in *American Men of Science* and several *Who's Who,* such as *Who's Who of Women, Education, Commerce and Industry,* and *Dictionary of International Biography.* She was elected to Sigma Xi, Phi Kappa Phi, Sigma Delta Epsilon, and International Platform Association, and is a member of American Society for Microbiology and Society for Industrial Microbiology. Her hobbies are duplicate bridge and nature study.

William A. Curby, M.S., is Head of the Sias Biophysics Research Unit of the Lahey Clinic Foundation and Director of the Alice Sias Memorial Laboratory in Brookline, Massachusetts.

He received his B.S. from Tufts University, Medford, Massachusetts, in 1950 and his M.S. from Tufts University in 1953. From 1953 to 1958, while a Research Assistant Professor at Tufts, he was sponsored by Tufts School of Dental Medicine to study a program involving course work at Massachusetts Institute of Technology, Tufts University School of Medicine, and Boston University School of Medicine for the purpose of putting physics, biology, and medical science into a combined discipline. From 1948 to 1960 Mr. Curby was a U.S. Public Health Fellow at Brandeis University, Waltham, Massachusetts, in a biophysics-biochemistry combination study program.

William Curby has been Principal Investigator for the Department of Defense, National Institute of Health, and Environmental Protection Agency research grants and contracts. For the past several years he has been a lecturer at Northeastern University, Boston, Massachusetts, at the Center for Continuing Education, teaching current state-of-the-art courses related to light energy measurement, photophysics, and advanced optics. He has been a consultant to industrial and federal groups in instrumentation and systems development.

Mr. Curby's research interests have been directed toward the study of real time analyses of biology and medical-physical phenomena. He is the holder of U.S. and foreign patents related to ultrafast-acting sensors for forces and for temperature, for microparticle capture and analysis, and for sensoring and processing methods related to automated monitoring and evaluating of real time changes in the physical and chemical characteristics of living cells. In addition to patents, Mr. Curby has published many papers, scientific articles, and reports on instrumentation and application of changes of cellular characteristics in response to applied physical and chemical stresses.

ACKNOWLEDGMENT

The considerable support of the Ames Division of Miles Laboratories, Elkhart, Indiana, during the preparation of this manuscript is deeply appreciated.

The courtesy and helpfulness of the personnel of several commercial firms who participated in the preparation of and reviewed the chapters related to their product is gratefully acknowledged, especially Dr. J. E. Ahnell of Johnston Laboratories, Inc., Dr. Phillip Wyatt of Science Spectrum, Inc., Mr. Robert C. Matteline of Vitek Systems, Inc., Dr. David C. Mayfield of Abbott Diagnostics Division, Dr. Paxton Cady of Bactomatic, Inc., and Dr. P. Blume of Good Samaritan Hospital, Portland, Oregon.

The author wishes to acknowledge the support and assistance of Sandy Pearlman of CRC Press in the preparation of this volume. The help of many in typing the manuscript is appreciated, especially Vicki Bilenker, Vicki Carpenter, Joan Chapman, Jann Charlston, Mary Drummond, and Gloria Froelich.

Dedicated to
my mother, Vida S. Gall,
and to
A.C.J.M.

TABLE OF CONTENTS

Chapter 1

INTRODUCTION

I. NEED FOR INSTRUMENTED SYSTEMS FOR MICROBIOLOGICAL ANALYSIS OF BODY FLUIDS

Remarks like, "You bug-hunters have progressed about three years beyond Pasteur!" have spurred microbiologists to think increasingly about improving the methods for microbiological analysis of body fluids. Their commitment has been strengthened when physicians explain that they do not send clinical specimens to the bacti lab because "the patients are either dead or gone home by the time the results are available." Clearly, better, speedier, microbiological analysis is needed.

II. GOALS FOR INSTRUMENTED SYSTEMS

What are the methods that require improvement? Briefly, a complete microbiological analysis using traditional methods usually consists of three major interrelated phases: (1) detection of pathogenic bacteria, sometimes coupled with enumeration, which takes a 16- to 24-hr incubation period to isolate bacterial colonies on agar media for enumeration and further testing, (2) antibiotic-susceptibility testing on the pure isolate by the Kirby-Bauer disc method or for Minimum Inhibitory Concentration (MIC), which takes another 24 hr, and (3) identification of the pure culture of predominating pathogens by differential selective media or antisera which may require 2 to 3 days after isolation. Thus, a complete analysis may take 3 to 4 days. Since the procedures are usually performed individually and manually, the results are often reported slowly and are sometimes inaccurate, as well as costly.

The problems have been stated. There is a need for more speed, accuracy, and lower cost in microbiological analysis. What has been done to achieve these goals?

In the late 1960's and continuing into the 1970's, simple instruments aimed at assisting one discrete part of the analysis were devised, mainly as labor-saving devices. These included media dispensers, plate streakers, colony counters, gram stainers, antibiotic-sensitivity readers, and microdilutors. These helped to relieve the tedium of some processes, speeded up the analysis to a limited extent, and by reducing the manual work load, may have cut costs. However, most of these devices have had little impact on the average microbiological lab. Perhaps the exception may be the microdilutors aimed at performing MIC tests. They seem to have a chance of challenging Kirby-Bauer's dominance in the field of antibiotic-susceptibility testing. These devices will be covered briefly in a later chapter.

III. HISTORY OF RAPID PROCEDURES USING KITS

Starting in the late 1960's and early 1970's, attempts to improve microbiological analysis got into full swing with the appearance of many products termed, "rapid methods". Many of these "systems" centered around kits which employed conventional metabolic tests to identify previously isolated bacterial cultures using paper strips, or miniature "test tubes", or containers with a preset battery of tests. Among these were Pathotec® (Warner-Lambert), R/B® (Corning), Enterotubes® and Oxiferm® (Roche), API® (Analytab Products, Inc.), and Minitek® (BBL). Most were aimed at identifying Enterobacteriaceae. These test kits often offer decreased incubation time, standardize the procedures, increasing accuracy, and, by combining many individual tests, often reduce the overall cost and time of identification. Recently, some of these kits have been backed up by computer analyses, offering probabilities rather than empirical identification. Studies on most of these kits have been reviewed recently[1-5] and are also the subject of numerous journal articles.

Whereas these kits give acceptable results in the identification of pure cultures, they still require a 16- to 24-hr incubation period for isolation, and there are manual procedures both in setting up and in reading the tests. Antibiotic-susceptibility tests must be performed as a separate procedure. Interpretation requires a skilled microbiologist and is based largely on color changes, some of which are difficult to read. Many of these test kits and simple instruments are widely used as diagnostic aids in clinical microbiology laboratories and are undeniably of value to the microbiologist, but there still is room for improvement.

IV. NEED FOR MORE SOPHISTICATED, DECISION-MAKING INSTRUMENTED SYSTEMS

Convinced that a better solution to the problem is needed, the microbiologists turned to the chemists and hematologists. They were once in the same situation as the struggling bacti lab, but now have thriving, profitable laboratories. Microbiologists asked them what to do. They responded in chorus, "Instrument and automate your laboratory procedures! Get some sophisticated decision-making instruments and automatic analyzer-recorder-computer systems into your labs!"

The microbiologists have apparently been persuaded that is the way to go, judging by the flurry of activity in the field of instrumented and automated microbiology. Fortunately, this decision by the microbiologists coincided with advances in technology in electronic engineering and computer science, making more sophisticated instruments feasible.

V. NEED TO UNDERSTAND THEORETICAL AND PRACTICAL ASPECTS OF INSTRUMENTED SYSTEMS

These circumstances have produced a confusing array of instruments using widely varying approaches for performing microbiological analyses. No less than a score of complex instrument systems using as many different approaches have been the subject of extensive research and development, mainly by commercial enterprises. Some of these systems are on, or are approaching introduction to, the market, while others have failed or are in a holding pattern awaiting further development. This burgeoning industry has created a group of scientists and engineers who need to know more about microbiological automation, both from the theoretical and practical standpoint, in order to design or evaluate instrumented systems. Attempts to consider these aspects are only recently starting to appear in the literature.[6]

VI. SCOPE OF COVERAGE IN MONOGRAPH

This monograph is intended to go more deeply into the theoretical and practical aspects of microbiological instrumentation, especially as related to sophisticated, decision-making instrument systems which are designed, developed, and applied specifically to routine microbiological analysis of body fluids. A detailed illustration of the application of these considerations to the actual development of one instrument system will be presented, followed by examples of various instrumented approaches. An evaluation will be made with regard to their adherence to the fundamental microbiological principles which lead to the attainment of the stated goals discussed for instrumented systems. Finally, a look at the potential of instrumented systems to extend knowledge in clinical microbiology will be included.

The material presented in this monograph is aimed at a diverse readership including clinical microbiologists and pathologists who want to keep abreast of developments in instrument systems and for students entering the field. Special emphasis is placed on the subject matter particularly pertinent to the efforts of biomedical engineers designing microbiological instrumentation and of microbiologists planning to add such systems to their laboratory, as well as government agents and legislators responsible for regulating diagnostic devices.

REFERENCES

1. **Kronish, D. P.,** Development of reagent-impregnated test strips for identification of microorganisms, in *New Approaches to the Identification of Microorganisms,* Hedén, G. C. and Illéni, T., Eds., John Wiley & Sons, New York, 1975, chap. 22.
2. **Gallien, R.,** Enterotube Roche — A rapid and accurate method for the identification of Enterobacteriaceae, in *New Approaches to the Identification of Microorganisms,* Hedén, G. C. and Illéni, T., Eds., John Wiley & Sons, New York 1975, chap. 23.
3. **Nord, C. E., Wadström, T., and Dahlbäck, A.,** Evaluation of different diagnostic kits for Enterobacteriaceae, in *New Approaches to the Identification of Microorganisms,* Héden, G. C. and Illéni, T., Eds., John Wiley & Sons, New York 1975, chap. 24.
4. **Moussa, R. S.,** Evaluation of the API, the Pathotec and the impoved Enterotube systems for the identification of Enterobacteriaceae, in *New Approaches to the Identification of Microorganisms,* Hedén, G. C. and Illéni, T., Eds., John Wiley & Sons, New York 1975, chap. 25.
5. **Mitruka, B. M.,** New methods of detection and identification of bacteria, in *Methods of Detection and Identification of Bacteria,* CRC Press, Cleveland, Ohio, 1976, 102.
6. **Curby, W. A. and Gall, L. S.,** Instrumental approach to microbiological analysis of body fluids, *Public Health Lab.,* 35, 118, 1977.

Chapter 2

THEORETICAL ASPECTS FOR AN AUTOMATED MICROBIOLOGICAL ANALYSIS SYSTEM

I. BACKGROUND OF MICROBIOLOGY THEORY DEVELOPMENT

The success of analytical chemistry instrument systems in clinical medicine laboratories has helped to change the attitudes of microbiologists who had thought that no instruments would be able to provide the diversity of information that comes only from decisions of trained experienced personnel making decisions based on the pattern of metabolites produced by a complex grouping of Petri plates and test tubes filled with different nutrient broths and gels. During the 1960s, there was a general feeling that, while an automated clinical microbiological analytical machine would be desirable, it would not be possible to develop because of the fact that biological phenomena vary in precision by ± 10%. Research workers finding differences in microbial growth characteristics from those in published works of others using organisms of the same genus and species had a simple explanation. They assumed that the differences were due to variations in strains. Several clinical laboratories using the Bauer-Kirby technique for microbial sensitivity to antibiotics will explain differences as due to use of a "modified" Bauer-Kirby method.

A. Need for Precise Information

1. Interactions of Engineers and Microbiologists

During the time of the original rocket probe launchings to the moon and to Mars, the National Aeronautics and Space Administration (NASA) had a "Planetary Quarantine Section" whose work involved advising the engineering and rocket-launching groups on how "sterile" each component of the rockets and rocket instruments had to be. Many heated debates took place between the engineers who wanted "a number" and the NASA biologists who could not guarantee 100% decontamination unless all microorganic life was destroyed. Data which would give a probability of decontamination were not available because no agreement was ever forthcoming from all of the biologists who were asked for an opinion. The debates did bring to focus an interesting set of attitudes of engineers towards the microbiologists and of the microbiologists towards the engineers. The engineers claimed that if they had the time, they could take a few biology books home for a week and get all the necessary information to solve the problem of how to define how biologically clean the rocket would have to be to provide a 99.9% probability of asepsis. The biologists were not accustomed to specifying in a rigorous manner the properties of a viable cell or population of cells. They felt that the engineers would be satisfied with any number, because engineers are trained to look up numbers in handbooks and to apply the numbers to physically and chemically precise circumstances. A basic reason for the entire debate on planetary quarantine can be better understood by considering how each participant defined viability of a microbial cell. Several biology texts define viability as the ability of a cell to reproduce itself and are satisfied with this as a complete definition. The definition is not complete, however, because it does not contain a time term. "If we wait an hour for a cell to divide and nothing has happened, is the cell not viable?", ask the engineers. The definition is inadequate to answer the question. If cells are put into proper growth conditions, and nothing has happened for 3 days, are the cells dead or alive?

After 3 days if we see bacterial colonies form, is it because the cells of interest have divided or is it contamination? We cannot be sure. If we do not know the identity of our original cells and cannot verify the identity of the new colonies, we can never be sure. Exobiologists are questioned on the possibility that bacterial populations which they have isolated from meteorites or from deep arctic ice may be laboratory contamination rather than populations from cells that are truly old or extraterrestrial in origin. There is presently no way to eliminate all doubt of the existence of contamination to the satisfaction of all of those who raise the question. In effect, this and other joint projects combining the thinking of engineers and microbiologists pinpointed the need for more precise definitions of microbiological phenomena.

2. Approach to Development of Precise Information by Chemists and Physicists

The development of bacteriology as a technology from the time of van Leeuwenhoek, Spallanzani, and Pasteur has not progressed as broadly as the developments of physics and chemistry. Two major reasons probably account for this fact. First, the applications of the originally described techniques were adequate to provide information essential to the immediate alleviation of sickness and death of humans. Efforts, thus, have been made to refine the original techniques, rather than to develop new ones. Second, the microorganisms and their effects on human cells could be seen under the visible-light microscope and could be described in illustrations or texts. In contrast, early physicists and chemists could not always see what was happening, but could only see certain end results. This forced evolution of the theories of the properties of molecules and atoms by physicists and chemists was done largely by observing phenomena and trying to deduce what must be happening and what was causing the happening. The descriptions of the phenomena had to be extensive and exact.

B. Development of Precise Information for Microbiology

The development of precise and exact definitions of observed natural phenomena is the key to the successful development of any instrumented system. Biology is no different in this respect from any other exact science. It is one of the purposes of this book to provide examples of how precise and accurate descriptions of biological phenomena can be used to develop theories or statements of conditions under which predictable changes in biological cell states or population states will come about.

The scientist must be an observer of nature. He must then define carefully, completely, and exactly what he observes. He has to define the terms and conditions under which he has made the observations. He must also vary these terms and conditions in an exact fashion and observe how these changes have affected his previous observations. He must make permanent records of what he has observed, and the records must be in a form which is clear to others having similar basic knowledge. These statements of condition and statements of reactions to defined changes of the condition can provide a basis for thought. From these thoughts can come statements of general relationships which can be used to establish theories involving combinations of general statements, or to broaden the limits and ranges of the conditions under which the general statement will always be true.

Microbiology can be put through the process of observation and definition just as any other scientific discipline. If enough conditions can be defined, it will be possible to make general predictions of reactions just as in the disciplines of physics and chemistry.

C. Application of Precise Information in Instrument Design

The precise knowledge of how a population of microorganisms will react under set conditions is essential to design a system which is really effective as an automated clinical microbiology analyzer. The theoretical knowledge must be reinforced by information obtained by its application to classical microbiological practice. This approach for designing instruments, which exploits theoretical knowledge substantiated by actual practice, is not new, but it has not been used in microbiology. Many of the microbiological analytical systems on or near the market have mainly depended on instrumentation of traditional microbiological analysis, considering only an end-point reaction, with little regard to the theoretical aspects of microbiology. As a consequence of this, the output of these systems has been chiefly empirical.

D. Development of Model

In designing a microbiological analytical system, it is first necessary to define the model which is going to be considered for building the logic for interpretation of the status of microbial cells.

Such a cell-state interpretation logic must be broad enough to satisfy all of the present clinical microbiology needs and, yet, be able to include new techniques as they appear.

After observing natural phenomena, a decision is made as to the best conditions for observing changes in the state of a population of growing cells. We selected as a model a population or populations of microorganisms growing in a fluid nutrient medium. Other models could have been chosen, however, but they did not offer as many easily definable properties. The final selection of a model by a designer is always strongly influenced by his ability to define conditions for predictable actions and reactions of the model. In general, the simple models are the best for design of state-of-being logic.

Once a model is chosen, rules which describe the conditions under which predictable actions can be obtained must be developed. The rules should cover the behavior characteristics of the model to a depth necessary to develop basic system-design logic. Just as in physics and chemistry, the development of an automated biologic-analysis system must be based upon biological principles and their corollaries. These act as the foundation for the system.

II. APPLICATION OF THEORY TO THE DEVELOPMENT OF AN AUTOMATED INSTRUMENT SYSTEM

A. Properties of Microorganisms Growing in Fluid Medium

To cover the scope of needs for developing an automated clinical microbiology analyzer, five rules were formulated based on the properties of microorganisms growing in a fluid medium. These will form a solid foundation for the development of a versatile analysis system, and the rules are stated in simple terminology which describes the general subject covered. They are the basis for the selection of a sensor transducer and for the methods of data collection and manipulation necessary to produce information needed for the treatment of such persons. The rules are (1) the Concentration Rule, (2) the Dilution Rule, (3) the Equilibrium Rule, (4) the Reaction Rule, and (5) the Dimension Rule. Each describes a property of microorganisms growing in a fluid nutrient medium.[1] The rules are stated in the following sections.

1. The Concentration Rule

When the minimal cell concentration needed for cell reproduction is equaled or exceeded, the initial lag interval following the inoculation from a stock bacterial population is independent of the number of organisms in the inoculum.

This fact allows the study of the growth of bacteria in a sample of body fluid regardless of the number of organisms in the sample, providing that the minimum growth-concentration limit is exceeded. In practical cases, this is usually less than 100 cells per mℓ of fluid growth medium. This rule also allows the use of an analysis system which can be based on fixed time intervals.

2. The Dilution Rule

A population of bacteria growing in log phase may be diluted with medium having exactly the same chemical, physical, and colligative properties as the initial growth medium without interrupting or altering the log-growth rate of the population.

This rule makes it possible to put continuity into the sampling procedure and also provides a method for obtaining many samples each identical to the rest. The continuity is established since only one diluted sample need be allowed to continue to grow while all the others are available for the experimental addition of chemicals, biologicals, or applied physical stresses. If more test samples are needed, the control sample is available for further dilution. It can be demonstrated in the laboratory that this subdilution procedure can keep a population of bacteria in log-phase growth literally for days or weeks. The rule also indicates that this methodology can be used for putting a growing bacterial population back into any degree of the lag phase, depending on how inexactly the medium properties of the diluent are made prior to the dilution.

3. The Equilibrium Rule

Any population of bacteria from a fluid source is in a potential growth-equilibrium state that is controlled by all other biological and chemical constituents contained along with it.

This rule allows dealing with a system that works directly with bacteria contained in body fluids without separation of different bacterial populations or separation from natural inhibitors or enhancers. To obtain information in the shortest possible time, or to work with microorganisms in a growth state as nearly like their growth state at the site from which they were taken, techniques which require subculturing or colony separation and reinoculation should be avoided.

4. The Reaction Rule

A bacterium in log-phase growth under fixed conditions will respond to a controlled applied stress or insult in a predictable manner.

This rule is important because it defines the property of a bacterial population in log-phase growth in a fluid medium which allows the comparison of antibiotic susceptibilities, minimum inhibitory doses, antigen-antibody reactions, and synergistic effects of antibacterial compounds on the bacterial populations of interest. A corollary to this, and to the Equilibrium Rule in combination with it, is that each bacterial population in a mixed culture will behave independently to the stress or insult. This will result in the establishment of a new growth equilibrium. Thus, if one bacterial population is completely inhibited in growth and another contained in the same fluid sample is not, the second will continue to grow. If the second population was being inhibited competitively by the first population, it will increase its growth rate because of the loss of the first population. Alternatively, if the first population was producing a growth-enhancing substance for the second population, the loss of the first population will decrease the growth rate of the second.

5. The Dimension Rule

All microorganisms from a common source cell and growing under the same conditions have exactly similar physical dimensions and metabolic activity at the instant following cellular division.

The uses of this property of a bacterial population growing in a fluid medium may not be as obvious as the applications of the other four rules. However, it is probably the most important in setting the operating and design limits on electronic equipment intended to monitor bacteria as they grow. It tells that the size of bacteria can be used as one parameter in any analysis scheme planned. It further tells that if the size-difference discrimination is good enough, there is an excellent chance to "see" different types of bacteria in a mixed-culture sample.[2] This rule also gives a bridge to classical optical microscopy and tells the conditions under which an accurate calibration of any electronic sensor which will be used in a clinical microbiology analysis system can be made. The fact that the metabolic activity can be predicted at the time of division tells us that we can get the most precise and predictable reaction to stresses which alter metabolic activity by picking the time when the greatest number of cells are actively dividing. This time is during log-phase growth.

6. Summary

These rules give us the ability to use and modify our model, to predict what will happen in special cases, and to set the requirements for selecting electronic components and circuit operating ranges. Protocols for designing logic algorithms for obtaining information that gives the same information as a classical clinical diagnosis method is possible. The ability to program a new diagnostic interpretive procedure into the microprocessor of a system keeps the system from developing untimely obsolescence. This is an important consideration when an investment of over $50,000 is possible and when a staff training and education program has been completed. The ability to update a system easily should be considered and acted upon in the development of the final approach to specifying the instrument capabilities and logic algorithms.

III. GENERAL SYSTEMS DESIGN

A. Approaches to Design

The design of instrument systems which report on changes in natural phenomena depends on the precision and predictability of the observed facts, as emphasized in the theoretical rules. This can be approached in two ways. One approach is to exploit a phenomenon which can be measured by some existing piece of apparatus. This approach is the most frequently used and can be very effective. The modifications of the output results of the phenomenon are usually done empirically, and the precision and accuracy are not always clear. The biggest weakness of this type of instrumented approach is in monitoring combinations of phenomena and in predicting the success of application of the instrumented system to previously untried combinations of operating conditions. That this approach is practicable in microbiology is attested to by the fact that all of the semiautomated clinical microbiology-monitoring systems of today are based upon this approach.

A second approach to the development of such a system that can be used for making interpretations about the relative amount and intensity of the changes is based upon the evolution of a list of requirements which the system must meet. The rules set the operating ranges and limits under which the system is precise and accurate and the conditions which must exist if the results are to have real meaning. Although no present system on the market today has been developed this way, a system using pulse-height analysis as an approach for sensing, enumeration, and monitoring of microorganism population states is discussed in detail to serve as an example of the application of this second approach (see Chapter 5).

B. Basic Concepts for Design

1. Sensitivity

a. Actual Sensitivity

There are several basic concepts which must be considered in instrument design. One of the major differences in the approach to the design of an instrument that will measure the presence of and the amounts of chemicals in the human body as contrasted with one which measures the presence of and amounts of bacteria in the human body is in the concept of sensitivity. Physical and chemical analytical systems are designed to measure the smallest possible amount of a substance or property of interest. The smallest precise and accurate quantity or property measurable by an instrument is defined as its sensitivity. We will define it as "actual sensitivity". To measure amounts of a chemical in a fluid medium, for example, which are below the actual sensitivity of the instrument, the instrument must be made more sensitive, usually at an increase in cost of the instrument, or the process must concentrate the chemical in the fluid by a known amount. If the amount of concentration is enough to make the final concentration of the chemical in the fluid greater than the sensitivity of the instrument, it can be measured and the appropriate calculations made to determine its original concentration in the unconcentrated fluid.

b. Potential Sensitivity

In dealing with microorganisms, new thinking must be developed. We must understand the ramifications of "potential sensitivity". One bacterial cell has the potential to produce tens of thousands of bacterial cells within a few hours. Instruments must be designed to operate under conditions which allow a microorganism to reproduce itself to a population level which can be unmistakenly measured. The sensitivity is exploited in this case if the conditions are set so that a small number of microorganisms will grow to a measurable level in a predictable amount of time whenever they are present. In studies of microorganisms contained in a fluid of the body, it is usually advisable to dilute the fluid rather than concentrate it in order to get the greatest potential sensitivity (exactly the opposite of the actual sensitivity case). This is because body fluids usually contain natural inhibitors. By diluting the sample, the concentration of an inhibitor is sometimes reduced to a level which is less than effective in stopping the reproduction of the microorganism population or populations in the fluid, but which has not reduced the bacterial population below the critical level for growth. For example, in the study of bacteria contained in human blood, some of the formed elements in blood contain very potent bacterial growth inhibitors. Lysozymes are contained in monocytes, for example. Care must be taken to separate the blood fluid which contains bacteria from the formed elements in the blood without releasing any element containing bacterial growth inhibitors. Any premeasurement manipulation of fluids containing bacteria which are of interest must be done in a way designed to preserve or increase the potential sensitivity of the sample. It is apparent, therefore, that time and physical manipulation are extremely important factors in the exploitation of a sample having potential sensitivity.

2. Measurement Parameter

A second important concept that must be considered in the design of instruments, which will be successful in detecting and measuring living cells, is that the system must be capable of measuring changes of the basic parameter that affects the sensor transducer. In this chapter, rules have been given which tell how a population of living cells can be expected to conduct itself in a fluid medium. All of these rules are based upon observations of microorganism populations that are growing in a predictable fashion.

We have observed that the only period of bacterial growth during which the change in population number is predictable over a useful interval of time is when the population is in log-phase growth. Therefore, all rules and techniques for interpretation of the samples containing microorganisms discussed in this chapter presuppose that the microorganism population is in log-phase growth and that no interpretation can be made from single-point data. All interpretations must be based on changes (first derivative) or rates of change (second derivative) of the parameters which are being monitored. Instruments which measure endpoints only, or parameters at steady state, are not able to take advantage of potential sensitivity since the time between levels of steady state or endpoints is often too long to provide data which are detailed and timely enough to make immediate clinical decisions.

IV. REQUIREMENTS FOR A FULLY AUTOMATED SYSTEM

A fully automated system is one which requires no human intervention at any time during its operating cyle and which, therefore, has special, more stringent requirements than a semiautomated system. These requirements must be determined and applied during the entire design and development of a fully automated system.

A. Ideal Automated System

An ideal clinical microbiological analysis system would give the results of detection, enumeration, identification, and graded antibiotic susceptibility within a few seconds, or minutes at most. Also included in the printout will be all information identifying the patient along with aids to interpretation. The output will be translated to a form which is understandable to the physician treating the case and will contain the results of any special tests he requires. No such ideal system exists presently, but it is technically possible to make one that meets all of the stipulations with the exception of the time to complete the analysis. If some manipulation of the sample prior to insertion in the machine is allowable, it will make the system much less elaborate. At this time, a system can be built that will do all of the aforementioned things, but the complete analysis will take a minimum of 2½ hr and a maximum of 4 hr. Some manual transfers from sample blades or from swabs into liquid will also be required at present, but these operations can be automatically manipulated when necessary by computer control.

B. Computer Functions

Such a system is possible because of the availability of computer microprocessors that are low in cost and high in programming options and data-manipulation speed.

1. Verification of Data

A basic rule for reliable automation of a system is that each data-collection step must be verified while the data are being taken and after each step in which data are merged, manipulated, or transferred. This is easily done by current microprocessing techniques. For example, when a sensor is measuring a sample, a check should be available which confirms that the sensor is operating properly and that only data which are generated while the sensor is operating correctly will be passed into memory for storage. Further, if the sampling interval or volume is not correct, the sample will be repeated. Before and after each sample is taken, the system should check itself to prove that all subsystems are functioning properly.

Several methods of proving that the data are valid are available. For example, if only specific distributions on an x,y plot are possible, a simple pattern-recognition program can be written which will check to confirm that the distribution obtained from a sample is received in memory properly. In any form of derivative analysis in which the function of the derivative is known, a mathematical form of the phenomenon can be solved in the microprocessor in faster than real time, starting with the data values at the beginning of the analysis cycle and providing data values to compare with those being obtained from the sensor as the sample data are collected during the cycle. If the expected value varies from the observed value by more than a predetermined amount, the sample must be reevaluated. If the values do not agree after three tries, preprogrammed remedial action must be started. If the system cannot correct the problem using its remedial-action options, it will shut down and call for help. Further reliability can be added by building redundant units within the system. Then, if a problem is encountered, a second unit is put into operation as soon as the system detects trouble in the first unit. If the system corrects the problem through programmed remedial action, the secondary unit can be switched out and the primary unit switched back in at the first available "switch-unit allowable" signal.

2. Decision Making

Pattern recognition analysis, function analysis, and data comparison and manipulation programs are also used to make decisions that human operators have been formerly required to make. Decision making is much more possible when derivative data are used. The determination of which decision path to take is handled in the microprocessor where every complex decision is broken down into a series of yes-no or either-or decisions. Tests for contamination of a sample by the previous sample, or tests for the existence of contamination of a batch of fluid needed in the analysis, must be included. Tests which eliminate the use of a fluid which is not appropriate to the particular procedure being done must also be included. If there is a date-of-use limit on any fluid or solid component which is to be used in the system, a test for that particular restriction must be built into the analysis routine.

3. Awareness Programming Circuit

All of these methods of checking for the validity of obtained data and the validity of internal operating processes come under the heading of "awareness" programming-circuit design. All systems which are anticipated to be made fully automatic must have the "awareness" capabilities designed into the system plan in the beginning, since it makes a difference in the selection of the sensor transducers that are to be used and, also, in how the samples will be presented to the system. It can be further stated that systems which rely on single sampling of each datum item are reliable only if a complete awareness cycle is designed into the system. Systems which make many samplings of each datum item are more reliable, but circuitry which checks for reasonableness of data and proper operation of the electronic circuits is still necessary.

4. Reference to Standards

In the absence of awareness circuitry, automatic systems must rely on the frequent submission of standardized calibration samples to check the proper operation of the system. The weakness of this approach is that it does not allow for remedy and retest of the sample at the time that the sample is being analyzed. Reference to calibration samples and electronic standards rather than use of awareness subsystems is the least expensive approach to validation of data obtained from a fully automated system in the short run. It is the most expensive over a long period of time or when many samples are processed.

C. Other Requirements

Operator safety is usually easier to achieve when a fully automated system is operating. However, the decontamination scheme and positive control of samples containing live microorganisms must be built into the system's physical design and into the control logic of the system. Verification of the proper operation of every decontamination cycle must also be included in the awareness circuitry. When samples are finished, they must be disposed of in a manner which is within the regulations of the municipal, state, and federal laws at the site of operation.

V. RELIABILITY CONSIDERATIONS

As previously stated, the system designed to provide a clinician with data which gives him timely information for the treatment of persons who have an infection related to the presence of bacteria in abnormal numbers or sites in the body must be highly reliable in its operation. The data must be in a form that the physician can use.

A. Multiple Parameter Analysis

Compound data will be subject to some usual biological variation. If a standardized approach that measures a single parameter is utilized, the biological variations may cause several exceptions to be encountered. To reduce the number of exceptions, a design can be considered in which the standard approach is changed under clearly defined circumstances to alternate standard approaches. The logic necessary to tell when to shift approaches and what approach to shift to is complex. The circuitry necessary to provide for extraordinary standard approaches is expensive and adds to the possibility of system failure. If it can be avoided, it is not desirable to design a system which requires extraordinary standard approaches to allow the instrument to handle all applications of clinical microbiology without exception.

We can get around the problem by taking advantage of the properties of the biological variation. Advantage can be taken of the increase in the probability of accuracy of interpretation of a fact through the use of multiparameter analysis. If decisions are made as to the existence of a fact based on two related, but exclusive, sets of data telling of the existence of a fact, and if it is known that the existence of either or both constitutes a positive confirmation, the chances of missing a positive reading are reduced by the product of the ratios of the individual ratios of the chance of missing an existing positive fact by either set of data items. The relationship holds for any number of individual, exclusive, data item sets. Thus, if we missed a positive fact 20 times out of 100 by either of two sampling methods, our chances of missing the fact would drop to 4 out of 100 if either of the two samples provided a positive existence of the fact. A third exclusive sampling would reduce our chances of missing the existence of a positive fact to less than one chance out of 100.

For practical application, it can be assumed that no biological variation encountered will be greater than ± 10%. If each sampling technique is better than 80% reliable, if four exclusive sampling parameters are used to make a sampling decision, and if any two are positive, the chances of missing the existence of a positive fact become less than one chance out of 100. The restriction that two positive parameter data sets must be seen in the sampling of four parameters cuts the possibility of recording a true negative as a positive. Therefore, a good rule to follow in designing a system which has to make decisions based on the existence or nonexistence of a fact should be based upon data sets having at least three degrees of freedom. Put in another way: whether or not a fact exists should be based upon four separate reduced data-item sets to prove the presence or absence of the fact. In the design of a system, priority is given to favoring the parameters which are known to give the smallest possibility of a false negative reading and to the parameters that miss the existence of a positive fact the least often. This improves the reliability even further.

All of the semiautomated systems presently on or near the market use only one parameter (data-item set) to make decisions as to the existence of a fact. This is probably their greatest weakness, and leads to the need to list several exceptions to their ability to detect particular microorganism populations.

VI. ANALYSIS OF MIXED POPULATIONS

Another special consideration should be taken into account before attempting to design a system which will detect and measure changes in a growing biological population. Again, this is involved in the application of the concept of potential sensitivity vs. actual sensitivity. In a potential-sensitivity case, the sensitivity is based upon development of an amount of material which is detectable above the lower threshold of the measuring ability of the instrument. In the case of growing populations of microorganisms, either the integrated total of the number of organisms in a representative volume of the sample can be recorded or a number relative to the population amount can be substituted. If mixed populations of cells are growing, and if the cells have some unique recognizable property, it is of great advantage to observe each cell in a representative volume of the sample and to record the numbers of cells based upon the presence or lack of recognizable properties. This allows for much more flexibility when studying mixed populations and is important when studying the effects of competition of one microorganism population on the growth of another population in the same fluid. The identification of microorganisms by serological techniques is also more efficacious if the system is set to observe microorganism populations on a cell-by-cell basis rather than by using relative numbers to express changes in the total population or mixed population of organisms contained in a given volume of sample.

VII. OTHER CONSIDERATIONS

A. Importance of Growth Phase

The system that is designed to take advantage of the potential speed of growth of a microorganism population into its initial log phase and to study the reaction of the population to antibiotics before changes in the phases of growth occur is highly desired. Systems that do not rely upon the plating and growth of colonies of microorganisms contained in the original body-fluid sample are preferred to those which do. All of these restrictions on the design of a system that will produce a broad-scope clinical microbiological analyzer can be established before any actual development program is begun. Many approaches to the design of equipment which meets these requirements are possible, however one which was designed to comply with them[3] is discussed in Chapter 5.

B. Frequency of Readings

Whatever system is used to monitor the growth cycle of a microorganism population, it is important that cell numbers and other parameters of interest are obtained often enough to note any changes in the growth pattern from the expected pattern for any particular microorganism. This means that the sampling interval must be short enough to recognize changes from the norm early and to see short-lived, reversible changes when they occur. A good rule of thumb is to make the minimum sampling interval approximately one twentieth of the time needed for the population to double.

For practical purposes, this is about 1 min. It follows that the time to obtain the sample and the time to analyze the sample and put the data from the analysis into storage together should not take longer than the minimum sampling interval. Analytical systems which are slow in collection and processing of samples will have minimum sampling intervals set by the minimum sampling and processing time of the system. It can be shown that if mathematic modeling systems are being used, integration errors can increase as the sampling interval increases.[1] It may not be necessary to sample as often as the machine is capable of, but the minimum sampling interval of about 1 min must be designed into the final system.

C. Summary

The requirements for designing a system which is capable of analyzing, detecting and obtaining data that will give physicians the information necessary to predict the course of an infection for treatment and to follow the results of their treatment can be summarized. They make up a set of instructions to designers of systems that hold true regardless of what sensing system or interpretation logic is planned for the system. The closer the instructions can be followed, the more versatile will be the resulting system as a clinical microbiology analyzer and the easier it will be to apply "awareness"-logic techniques to the control of a fully automated system.

VIII. SELECTION OF A SENSOR

The choice of the sensor requires the most consideration given any component as this is the heart of the instrument system. It is the component which varies the most with the different systems on the market or under development. An in-depth knowledge of the limitations, capabilities, and operating ranges of the sensor being considered for selection to satisfy a particular job must exist. The selection of a sensor whose operating characteristics make it inappropriate to the clear production of desired information is the usual cause for system weakness and early obsolescence. An understanding of the principles by which the systems presently marketed or under development work will be instructive in helping to see how the systems interpret the state of conditions of microorganisms detected in the test samples submitted to them.

A. Various Sensors

There are several different sensors used in microbiology-analysis instruments. One uses a radioactivity detector, and one measures electrical impedance. The others detect changes in some form of visible-light energy. Before discussing the principles under which each system determines the state of condition of microorganisms in test samples, we shall restate the process of selecting a sensor transducer. The sensor must be able to report the state of condition of microorganisms unequivocably by measuring a distinguishing characteristic of the microorganisms which can be obtained for no other combination of circumstances or in any other state of condition of the population.

1. Bactec®

An approach to the consideration of selecting a sensor-detector principle which has not been discussed has been applied to the Bactec® system. The concept utilizes the fact that system sensitivity does not have to be as great if one can be certain that the signal being measured can come from one and only one set of conditions. If one can also be certain that the signal which will meet the set conditions can only be generated when an action which is to be monitored has taken place, the probability of making a false reading is greatly reduced. The certainty of sensing a phenomenon which is based upon an impressed action is always higher than it is with a detector with a sensor which passively observes a natural action. Action-reaction concepts take advantage of the Reaction Rule when applied to microbiology. In the case of Bactec®, the action is the timely addition of a stable metabolite containing radioactive ^{14}C to a sample which is questioned as to the possible containment of live microorganisms. If microorganisms are present and able to use the radioactively labeled nutrient in their metabolic activity, some of the radioactive carbon will eventually be expelled from the microorganism as a metabolic end product, primarily $^{14}CO_2$ gas. If a sensor to the beta radiation given off by the $^{14}CO_2$ is set so that only radioactive gas can be measured, and that absolutely no measurement of the beta energy from the liquid containing the radioactive-substrate nutrient can be seen by the sensor, then only when the nutrient substance is oxidized by cellular metabolism will the sensor see beta energy. In the time frame allowed for detection after introduction of the radioactive metabolite, the presence of detectable radioactive gas from the sample can only mean that microorganisms are growing in the sample (refer to the Concentration Rule and to the Equilibrium Rule). Because there is high certainty that the presence of $^{14}CO_2$ can be taken as an indication of a positive microorganism-containing sample, very little $^{14}CO_2$ has to be present. There is no other way under the particular set of conditions that the signal could have been generated, and thus, the number of microorganisms contained within the sample can be low. The microorganism-detection sensitivity of the system is, therefore, high. The particular problems with this approach include decomposition of the radioactive metabolite and the continually increasing restrictions placed upon the storage, use, and disposal of radioactive isotopes. The Bactec® sensor-detector principle also does not make possible the separate sensing of mixed populations of microorganisms or the consideration of bacteria on a cell-by-cell basis.

2. Bactometer®

One other system relies on a signal which can be generated only under a unique set of conditions. It utilizes a passive rather than an active detector-sensor concept. The Bactometer® series of instruments measure changes in electrical impedance by passing an alternating voltage through electrodes set into a multicell cuvette into the medium containing bacteria and out through a single reference electrode. Originally, each cuvette cell had its own reference electrode, and all reference measurements were made for each cell. The system is improved by the fact that all multicell cuvettes are now disposable. Thus, the electrodes are used only once. A repeat use of the electrodes requires an involved and lengthy cleaning procedure. As in any conductance-measuring system, the surface of the electrodes are of critical importance, and the time necessary to prepare a surface can be more complex and uncertain than the test procedure. Impedance changes as precise measures of bacterial growth have been successful in laboratory studies, however the reduction to practice of a system which will operate in a predicted manner in all cases has been made possible because all growth cassettes are exactly identical in electrical and physical characteristics. Thus, the quality control and adherence to close tolerances for the preparation and assembly of the physical materials used in the system are of critical importance. Here again, the detector-sensor principle is such that total combined populations of microorganisms can be seen with no chance of separating populations in mixed samples or evaluation of changes taking place in microorganisms on a cell-by-cell basis.

3. Sensors Using Visible Light — Differential III®, AMS®, Autobac 1®, MS-2®

The art of measurement of light energy emitted from a source is called sensitometry (Definition 1). To confuse the issue, sensitometry is also used to mean the art of generation of uniform exposures (exposure = intensity × time) of light energy (Definition 2). Science Spectrum Differential Series® use a special form of sensitometry (by Definition 1). The McDonnell-Douglas AutoMicrobic System (AMS)® and Pfizer Diagnostic's Autobac® series use different forms of densitometry. Densitometry is defined as the measurement of the amount of light-intensity reduction as a beam passes through some form of light attenuator. There are several ways that light may be attenuated. It can be absorbed, reflected, scattered, refracted, polarized, or blocked by opaque material. When a beam of white light passes through an attenuator, it is separated into a specular portion and a diffuse portion. The specular portion is what is left of the light beam passing in a straight line through the attenuator. It is reduced mainly through absorption and opaque blockage. The turbidity measurements made by the MS-2® measure light attenuation of this type. If one moves a photo detector which is sensitive to the wavelength spectrum of the light source so that it is out of the direct beam passing through the attenuator, the intensity will drop, but it will usually not fall to zero. This portion of the light cone emerging from the attenuator is called the diffuse portion of the light. The diffuse portion is made up primarily of scattered light, refracted light, partially polarized light in some cases, and light that is reflected by material in the attenuator as well as surfaces of the attenuator (depending upon the thickness). Some of the light emerging in the diffuse portion of the emission hemisphere is absorbed along its altered path through the attenuator. The hemisphere of emitted light, if completely diffused, can be made to be uniform at all angles starting at the center axis of the incoming beam and moving in any direction until the sensor is normal to the beam axis (90°). The AMS® systems measure the attenuation of the specular portion of a white (all visible wavelengths) light beam passing through selected growth media at different times during the incubation of bacteria cultured from the human body. They measure turbidity in which a small amount of the light attenuation results from opaque blockage and a large amount from scatter. If the bacteria in the light path are in a medium containing a colored dye, and if they metabolize the dye, a change in absorption will also cause a change in the attenuation. The Autobac® systems measure white light passing through 4° of arc whose center is at 35° away from the center axis of the input light beam. The diffuse portion of the emerging light at this angle is made up largely of scattered and surface-reflected light and is reduced by the absorption along the extended light path through the attenuator. All three systems measure the total of all light emerging from the attenuator regardless of the cause of the attenuation. The Science Spectrum Differential Series® takes advantage of the principle of Oster[5] (see also 8.B.1) and measures the amount of scatter of a beam of monochromatic, vertically polarized, coherent light (laser generated) at several angles to establish a scatter profile for bacterial particles contained in a fluid set in the path of the laser beam. Since scatter is wavelength dependent, it is possible when using monochromatic light to distinguish between an increase or decrease in the number of organisms in a fluid and a change in the shape of organisms when the number of organisms is not changing. The three densitometric approaches cannot make this distinction, and again, all of the four can sense only total microorganisms rather than separate populations present in the sample. Science Spectrum does make an analyzer which can inspect a single bacterium by holding it in an electric field and shining a laser on it. It is a research model at the present time.

B. Factors in the Selection of a Sensor

1. Flaws in Visible-Light Sensing

In a preliminary analysis of the sensing operating principles, one looks for transducers which have high sensitivity and low cost, are safe, and the output of which is simple to transfer into a form easily handled in the data-manipulation portion of the system.

On first evaluation, it appears that a sensor capable of measuring visible light will be the best. However, to fulfill analysis requirements, we need to evaluate bacteria on a cell-by-cell basis with an ability to differentiate changes in size from changes in shape. Two important weaknesses of a visible light system appeared, and these are important enough to cause us to consider an alternate sensor for a microbiological analysis system. The first is most important. In reviewing the work of Oster,[5] we became aware of the fact that as the average diameter of particles which are measured in a light path becomes less than 10 × the wavelength of the light in the path, the ratio of the amount of forward scatter to back scatter changes in different ways depending upon the shape of the particles in the light path. The wavelength of IR light which displays visible-light physical properties is between 0.7 and 1 μm. (Oster uses the ratio of forward 135° to back 45° from the light-path axes.) Visible light has a wavelength range from approximately 0.35 through 0.7 μm. The bacteria of interest vary in average diameter from 0.2 μm to approximately 3.5 μm. As they grow their shapes usually change, thus making it impossible to obtain unequivocal data regarding whether changes detected by the sensor are due to changes exclusively in size or shape or whether the changes are due to some combination of the two. Shortening the wavelength of the measuring light to obviate the problem is blocked by another physical phenomenon — the increase in general scattering of the light to an extent that size vs. shape measurements are impossible due to the increase in total background scatter. As the wavelength is shortened, the energy of the light increases, and this increase in energy increases the scatter exponentially by the fourth power (Rayleigh's law of scattering). This makes the use of blue or violet light ineffective for the study of bacterial growth on a cell-by-cell basis. Experiments carried out in our laboratory have proven that Oster's predictions apply to bacteria. The second reason for not using visible-light variation as the sensing principle was that not enough information could be obtained from the alteration of light path to give an extra degree of freedom in our multiparameter data-interpretation requirement without going to a very elaborate and expensive preparation of the light and recovery of the light by multiple sensors.

C. Selection of PHA as the Sensor Scheme

Our choice for a sensor was pulse-height analysis which involves the analysis of the height and shape of a voltage pulse generated when a bacterium drawn through an aperture alters the fixed current-flux density in the aperture.

1. Basis for PHA Selection

On balance, we found that even with the drawback of aperture plugging, modification of the principle used for sensing particles in the Coulter Counter®* and Celloscope®** would fit our requirements. Most important, in the pulse-height sensing system, the shape and size characteristics of bacteria can be separated in the pulse generated by each bacterium altering a current flux, and secondly, signal-to-noise ratios can be increased statistically because of the uniformity of the bacteria being analyzed. The problem of coincidence can be controlled if the sample concentrations are kept below easily achieved concentrations.[4] With an eye towards observing viruses, the principle of drawing a particle through an aperture containing a fixed current flux and altering the density of the current flux proportionally to the cross section of the particle normal to the flow axis of the particle appealed to us in a special way because, in

* Coulter Corporation, Hialeah, Fla.

** Particle Data, Inc., 111 Hahn St., Elmhurst, Ill.

theory, there is no lower limit to the size of particle which can be measured using the principle. As long as the electronic noise of the measuring system is kept lower than the voltage generated by the passage of particles through an aperture, and as long as the current and particle-carrying fluid is kept clear of contaminating particles in the same size range, it is possible to detect and get information about the particles of interest to us.

IX. GENERAL INSTRUCTIONS FOR THE DESIGN OF A FULLY AUTOMATED, WIDE-APPLICATION, CLINICAL MICROBIOLOGY ANALYZER SYSTEM

A. List of General Instructions

Following the reasoning of the previous discussion, general instructions for the design of a fully automated microbiology analysis system are presented.

1. The system should measure the microorganism on a cell-by-cell basis rather than on an integrated total-population basis.
2. The sensing principle used should be unequivocal in the identification of distinguishing characteristics of the cells studied and in separating them from sample background noise.
3. Any approach which favors the exploitation of potential sensitivity should be given high priority when considering incorporation into the system design.
4. All interpretations of the state of being of the microorganism population must be made from data obtained when the cells in the sample are in the log phase of growth.
5. The system logic must be limited to the consideration of detection and monitoring the growth of microorganisms taken directly from the human body. No preinoculation and growth on intermediate media or physical separation of mixed populations should be accepted in the design protocol.
6. The output from the sensor transducer should have at least one operating characteristic which can be monitored to be certain that it is functioning properly.
7. High selectivity should be worked for.
8. If possible, four related, but exclusive, data items should be collected at each sampling time, and they should be used in all decision-logic calculations which determine the state of the population or populations in the sample. If four data items are not possible, three will be taken as an absolute minimum to be considered in any design approach.
9. Only systems which are designed to monitor near-real-time or real-time changes in the populations of microorganisms in the sample should be used. No end-point data or single-measurement data should be considered in the interpretation-logic protocol.
10. The system should be capable of obtaining all data for each sample and processing these data within 1 min.
11. A method for decontamination and testing for completeness of decontamination must be considered in the system.
12. The output from the system must be in a form which is easily understandable by clinicians and technicians schooled in classical medical microbiology techniques.

X. PRACTICAL APPLICATIONS FOR AUTOMATED MICROBIOLOGY ANALYSIS SYSTEMS

Following are examples and a discussion of techniques which demonstrate the potential power of a high temporal resolution, high cell-selectivity system.

The power of a system which is capable of observing and recording characteristic properties of representative cells in a population of cells once a minute may not be obvious to one who has not had time to consider the subject fully. Perhaps the easiest way to demonstrate this potential is to list several applications of such a system related to clinical microbiology. All of the examples listed have been, prior to this time, either difficult or too expensive in time and money to make them practical for routine use. These applications are easy for a real-time-analysis system whose design has taken into account the theoretical considerations discussed in this chapter. They should form the basis for understanding some of the potential ability of future clinical microbiology instrumentation. The examples given have all been tried and found to be easily able to be accomplished using the system concept which has been developed around a pulse-height-analysis sensor. The theoretical rules and considerations which have been discussed in this chapter, however, clearly show that any sensor system which can comply with the stated system requirements can be just as effective.

The following lists applications which are possible with real time monitoring of bacteria in log-phase growth, but are difficult or impossible using classical microbiology techniques.

A. Bioassay Techniques

1. Therapeutic-agent levels in body fluid can be determined by alteration of normal growth characteristics of a population of a specific test bacterium.
2. Alteration of specific auxotrophic-test-bacteria growth characteristics by the addition of mutagenic compounds or promutagenic compounds plus naturally occurring cellular mediator compounds (Ames test) can be done.
3. Determination of phagocytic potential of granulocytes to test populations of bacteria before and after a granulocyte donor has entered therapeutic treatment can be done.
4. Determination of the critical amounts of, and importance of, chemical compounds needed per cell to provide a complete growth medium for a particular bacterial population can be done.
5. Determination of the "wildness" of a specific bacterial population by its ability to react to a specific antiserum and a review of its growth characteristics can be performed.
6. Direct bioassay techniques for monitoring quality levels of pharmaceuticals are possible.

B. Viability and Infectivity Potential

1. Determination of comparative growth potentials of different populations of the same or different microorganisms before and after a controlled amount of physical or chemical stress insult has been applied to each population.

C. Quality Control

1. Development, monitoring, and quality control of test and calibration populations of microorganisms on the basis of specific, reproducible, and unique population-growth characteristics is possible.
2. Development and monitoring of the specificity and titer of antisera to bacterial strains is possible.

D. Modeling

1. Modeling the course of complex infections and determining and testing of growth countermeasures in the model is possible.
2. Modeling the results of sequential and/or simultaneous combinations of antibacterial growth agents in the study of control of chronic infections is possible.
3. Modeling the interactions of two or more bacterial populations competing for growth in a particular medium when the exact number of cells in every population is set at the beginning of the test can be done.
4. Modeling the growth, detoxification, and biodegradation capabilities of microorganisms to be used in industrial pollution control processes can be done.
5. Modeling growth characteristics and "scale factor" inoculation requirements for large-scale industrial processes using microorganisms is possible.

E. Identification

1. Identification of microorganism populations by strain using the addition of specific antibodies to test microorganism populations is possible.
2. Identification of the presence of particular resistive bacteria strains for particular hospitals is possible.
3. Identification and monitoring of sources and vectors of resistant microorganisms is possible.

F. Clinical Diagnosis

1. Determination of whether the number of particles in a body fluid is normal or is abnormally high is possible.
2. Determination of the effectiveness of antibiotic therapy by observing the growth potential and characteristics of microorganisms contained in particular body fluids of a patient receiving the antibiotic therapy is possible.

G. Summary

All of these applications are possible because the microorganisms, cells, and other viable particulates are set into an equilibrium state which can be quickly characterized as to numbers, sizes, and shapes. The system can, if desired, be stressed by applying controlled amounts of physical or chemical insults, and the reactions to the stresses can be observed and interpreted. The applications given as examples should help to demonstrate the concept, and indicate the power, of dynamic analysis when applied to biological systems.

REFERENCES

1. **Curby, W. A. and Gall, L. S.,** Instrumental approach to microbiological analysis of body fluids, *Public Health Lab.*, 35, 118, 1977.
2. **Curby, W. A.,** Microparticle Analysis, U.S. Patent 3,919,050, 1975.
3. **Curby, W. A.,** Dynamic Method of Identifying Microbes and Evaluating Anti-Microbial Processes, U.S. Patent 3,804,720, 1974.
4. **Curby, W. A., Swanton, E. M., and Lind, H. E.,** Electrical counting characteristics of several equivolume micro-organisms, *J. Gen. Microbiol.*, 32, 33, 1963.
5. **Oster, G.,** The scattering of light and its applications to chemistry, *Chem. Rev.*, 43, 319, 1948.

Chapter 3

PRACTICAL ASPECTS OF INSTRUMENTED SYSTEMS FOR MICROBIOLOGICAL ANALYSIS

I. INTRODUCTION

The theoretical aspects of the microbiological analysis of body fluids discussed in the previous section are of vital importance in designing or evaluating the instrumented system, as violations of the fundamental theories will result in invalid implementation of the chosen approach. But what about the practical aspects of microbiological analyses that are of concern both in designing and evaluating the instrumented system? This section will present the practical biological, engineering, monetary, educational, and service considerations intended to aid the potential user in evaluating how the instrumented system meets the goals of speed, accuracy, and reasonable cost, and discuss the engineer's special problems in designing an instrument system for analysis of potentially hazardous living organisms.

II. BIOLOGICAL CONSIDERATIONS

With respect to sophisticated instruments, several biological facets of microbiolgical analyses enter into design and evaluation decisions, including factors affecting test cultures, test methods and data measurements, reduction, and interpretation.

A. Factors Affecting Test Cultures

There are several factors affecting test cultures, including sample preparation, inoculum, and phase of growth.

1. Sample Preparation

The first step in the microbiological analysis is the sample preparation. Specimens received in the laboratory frequently need some manipulations such as dilution or concentration to ensure a suitable inoculum for a microbiological analysis. Such manipulation of the sample introduces some alteration of the environment which may delay log growth as well as cause a possible source of contamination. Therefore, all specimens should be prepared as rapidly and as aseptically as possible, with simple methods to be preferred.

During sample preparation, care must be taken to eliminate any substance that would interfere with the test. This will vary with the different analytical approaches chosen. For example, contaminating particles are undesirable in tests involving light scattering or pulse height analysis, especially the latter. Conflicting sources of bioluminescence may cause inaccuracies in procedures measuring luminescence, and radioactive material in radiometric procedures. When patients are receiving antibiotics, the specimen should be diluted as much as possible before tests for detection are done and in analyses where inhibitory substances are used to aid in detection and identification it would seem to be advisable to remove these inhibitors before performing an antibiotic susceptibility which is essentially a test of inhibition of growth. Leaving the inhibitor in the susceptibility test medium means testing an inhibitory reaction in the presence of an extra inhibitor not related to treatment.

2. Inoculum

a. Size of Inoculum

Controlling of the size of the inoculum is an important factor in instrumented microbiological analyses.[1] This control is usually effected by dilution or concentration of the specimen or test culture during preparation. In short duration rapid tests, too low an inoculum may fail to generate enough baceteria to reach the detection sensitivity of the instrument in a reasonable time period. On the other hand, too many bacteria may overwhelm the maximum detection mechanism of the instrument or of the test substance, as in the case of antibiotics. This factor is recognized in the standardized disc-agar plate method, such as Kirby-Bauer, where the inoculum concentration is carefully standardized.

b. Stage of Growth

The stage of growth of the inoculum may also influence the test results. A culture in lag or stationary phase may take too long to revive or be too sensitve or insensitive to be suitable for identification or susceptibility tests on instruments designed for rapid methods. With those instrumented approaches capable of performing detection, the use of the just-detected culture in rapid log phase is ideal for further tests. If a rapid log phase culture is unavailable, the culture must be gotten into log growth before starting the test to ensure the most accurate results. Cultures in different phases of growth may respond differently to antibiotics and in those approaches that use serological agglutination techniques for identification, the stage of growth is most important, as that influences the antigenic composition and combining site availability of bacteria especially with respect to the "O" antigen and capsular (K) antigen.

c. Clumping of Bacteria

The clumping,chaining, or grouping of the bacteria in the inoculum may result in "false" positives when the results of certain analytical approaches are compared with data obtained from plate counts. It is well known that plate counts which are used as a standard to judge automated results are too low and do not express the true bacterial population when the bacteria in the inoculum are not present as single cells since each clump of several bacteria produces only one colony. However, these clumped bacteria each exert an independent metabolism and produce metabolic end-products in proportion to their number regardless of their physical separation. Thus, methods which measure end-products of metabolism such as radiometric assays, luminescence measurement, and impedance will appear to indicate too many bacteria when compared to traditional plate counts whenever clumping or chaining is present. On the other hand, methods that depend on particle counting may yield too low a result, when clumping or chaining is a factor, especially as growth proceeds, as each clump or chain gives only one particle count, unless separated by shaking.

d. Mixed vs. Pure Cultures as Inoculum

Another important aspect with regard to the inoculum of the test culture is whether it is a pure culture of one type of bacterium or a mixture of two or more types of organisms.[2] Classical procedures require the test cultures to be pure, especially when identifying the bacteria by metabolism, as each bacterium is characterized by a distinctive pattern based on several metabolic tests. Mixed cultures will produce additive metabolic results from both types of bacteria, precluding the establishment of a metabolic pattern related to either one of the bacteria of the mixture. Therefore, there is a sound reason for purifying bacterial cultures to be identified by metabolic tests.

The question of mixed vs. pure cultures is of major importance in instrumented microbiological analysis. Certain instrumented approaches offer the possibility of doing the complete microbiological analysis — detection, identification, and antibiotic susceptibility testing of predominant pathogens — on the original specimen of body fluid without previous isolation and purification procedures. This is possible as the bacterial culture which grows during the detection phase of the analysis, whether mixed or pure, can be tested for antibiotic susceptibility and identification by methods inherent in the analytical approach which allows each organism to express itself individually. These methods offer the greatest potential for becoming truly rapid methods of microbiological analysis, since the overnight incubation needed for purification is eliminated. It should be pointed out that the mixed culture studies have been done mainly on urine specimens which are not as complex as some other types of specimens, such as throat or sputum cultures.

Also based on the interaction of microorganisms, antibiotic susceptibility testing on mixed cultures may produce a better treatment guide than classical procedures. The concept of pure test cultures is extended to antibiotic susceptibility testing. Is this wise? Many pathogens co-exist in the body with other bacteria, both pathogenic and commensals and these mixtures of bacteria interact. One bacterial type may be stimulated or inhibited by the presence of another type of organism, and these mixtures of bacteria interact.

These interactions are lost when the two types of bacteria are separated, as in purification. This may affect the results of tests aimed at treating a patient harboring a mixture of bacteria at the site of the pathology. For example, Organism A, found in mixed culture with Organism B, may be somewhat inhibited by a substance produced by Organism B. During classical purification the organisms are separated and tested for their susceptibility to various antibiotics. The results of the Kirby-Bauer test may show that Organism A is resistant to Antibiotic XY. Yet test results when available show the patient responding well. Why? Perhaps the combined inhibition of Organism B and the Antibiotic XY in the body was enough to inhibit Organism A and to allow the body defenses to overcome Organism A — i.e., allow Antibiotic XY to adequately treat the infection. Or the reverse could occur if Organism A was stimulated by a substance produced by Organism B. Upon purification Organism A, lacking the simulatory substance produced by B, might well show a susceptibility to Antibiotic YZ, but in the body, treatment of Organism A with YZ antibiotic might fail. Thus, a method of antibiotic susceptibility testing which allows the interaction of the organisms present as mixed cultures during the test period, may give results more representative of those in the body.

Certain microbiologists object to this conclusion stating that in mixed cultures in vitro not all of the bacterial types in the mixture will grow at the same rate as in the body and soon the fastest growing bacterium will predominate in vitro regardless of its relative predominance in the body. This certainly is true when cultures are allowed to grow for 16 to 24 hr, representing many generation times. However, with most of the rapid procedures, detection is accomplished after only two or three generations when the proportion of the various organisms is probably nearer to that occurring in the body. Thus, tests performed on the mixed cultures by rapid procedures may not vary too much in the predominance of member bacteria and may more nearly reflect conditions in the body, provided the chosen approach can distinguish which organism in the mixture is susceptible or resistant to each antibiotic.

e. Similar Colonies

Some of the instrumented procedures for susceptibility testing direct that the inoculum be prepared by picking several "similar" colonies. As similar appearing bacterial colonies may not always represent the same type of organism, the picking of several colonies may introduce the risk of having mixed culture inoculum, which is undesirable with most instrumented systems. This method of picking several colonies is also used in traditional microbiological techniques, but these procedures usually are based on the subsequent growth of the test cultures on an agar surface (Kirby-Bauer), where the presence of a mixture can be detected when different susceptibilities cause the resistant organisms to make its presence known. In most instrumented procedures such a mixture might well go undetected.

3. Phase of Growth

a. Fundamental Growth Patterns

Most rapid instrumented procedures for microbiological analyses of body fluids are performed on actively growing cultures in broth medium. The following brief review of the fundamental in vitro growth patterns of bacterial cultures lays a foundation for understanding the application of the theoretical biological rules stated in the previous section.

In an adequate growth medium bacterial cultures pass through four recognized stages of growth,[3] (1) lag phase when the culture is orienting itself to a new environment, but showing little reproduction, (2) the logarithmic phase when the culture is dividing rapidly in response to the new environment to which it has adjusted and during which time it is most uniform in its morphology and metabolic activities, (3) the stationary phase when the bacterial population is dividing slowly, producing an equilibrium between new cells formed and bacteria that are dying, and (4) finally the death phase when there is no cell division and the bacteria begin to disintegrate or autolyze.

The most reproducible, reliable data occur when the culture is growing logarithmically as the rapid division results in uniform cells in contrast to the irregular morphology and physiology of the bacteria in lag phase and the mixture of old and new cells in the stationary phase. For this reason a rapid automated procedure requires that a test culture be in log phase to make optimumly accurate measurements. Fortunately, rapid test methods can often be completed within the period of uniform log growth, and this outstanding advantage should be recognized when evaluating instrumented systems of microbiological analysis.

b. Log Growth Phase

The validity of the data then depends on attaining and maintaining the culture in the rapid log phase of growth. How is this accomplished? To get a culture into log growth, environmental conditions, which may vary with different bacterial types, play an important role. Fortunately, most common bacterial pathogens respond to a similar environment which includes 35 to 37°C temperature, a rich nutrient broth, and gentle agitation, and these favorable conditions should be provided by the instrumented system.

When presented with these favorable conditions, bacterial cultures will go through the lag phase period of adjustment and then start to divide logarithmically, continuing in log growth as long as conditions remain favorable. When the food supply is gone, metabolic products pile up, or there is a change in temperature, for example, the culture will fall out of log growth. The culture will resume log growth following a lag period after favorable conditions are restored; the length of the lag period is often related to the severity of the deviation from the favorable conditions. Therefore, to ensure accurate data, it is important to maintain the culture in the log phase of growth by providing a favorable environment at all times and when transferring or diluting the log culture from one environment into another to make sure that the culture is placed in the same conditions from which it is being transferred. This has several advantages. It allows indefinite extension of the experimental period without having to build up the population to log phase again, and provides an unlimited supply of inoculum from similar samples, all in the same phase of growth.

There is another possible advantage to using test cultures in the log phase of growth. Bacteria causing an infection in the body are probably in the rapid phase of growth, for if the infecting organism were growing slowly as in the lag or stationary phases the infection would be contained by the body defenses. Using rapid methods, the test culture will be in the log phase of growth when it most resembles the bacterium in the infectious process, and test conditions such as antibiotics imposed on these cultures may more nearly reflect the reaction occurring in the body.

In the detection of pathogenic microorganisms the initial phases of the test will probably be performed on a specimen with bacteria in the lag phase of growth, due to their recent change of environment from the body to in vitro conditions. These specimens should be placed into a favorable environment to encourage attainment of rapid log growth as soon as possible and then handled in a manner to maintain them in log growth. This is particularly important during identification and antibiotic susceptibility procedures, as different test conditions are imposed on aliquots of the same culture, all of which must be kept growing in a uniform manner.

The test culture (or specimen) becomes an integral part of the instrumented analytical system as soon as the specimen or inoculum is introduced into the instrument. Thus, the requirement that the test culture be in the active log phase of growth during the test period influences the entire test system and procedure. The log phase culture is the most uniform stage of growth and thus offers a more standard inoculum. Since one of the advantages of instrumented analysis is the standardization and control of the test procedure to give uniformity to the test procedure, the uniformity of the log phase inoculum is a plus value.

c. Source of the Test Culture

The source of the test culture, whether grown directly from the fresh clinical specimen, from an incubated clinical specimen, or a stock culture exerts a marked influence on the culture, both with respect to morphology and physiology. Transferring or even holding a culture on laboratory media induces changes so that the bacterial population is different from that in the clinical specimen, which the doctor is trying to treat. The advantage of the new rapid methods is that the pathogens are tested within a few hours of detection rather than a day or more later and are more nearly like the bacteria in the body, thus forming the basis for better treatment guidance. Full advantage should be taken of testing these less altered cultures by using rapid instrumented analyses.

To summarize briefly, for optimum results use a simple preparation of the test culture derived from the original specimen or a fresh primary isolate yielding an inoculum of reasonable concentration in rapid log growth.

B. Factors Affecting Test Methods

The methods chosen for instrumented microbiological analysis are subject to several practical biological considerations. These include provision of an acceptable test environment, employment of suitable experimental methods, and maintenance of adequate control over biological reactions by procedures that are compatible with instrumented capabilities. The methods developed to implement each of the various approaches will be different, but there is some commonality in the biological aspects of most approaches and certain biological requirements will place restrictions on the instrumented system.

1. Test Environment

All methods must provide a suitable environment for the growth of the test culture.

For example, most clinical pathogens grow well at body temperature so the incubation temperature can be maintained at about 35 to 37°C. Certain procedures may also require a steady temperature to ensure accuracy of the readings or optimum reaction of the reagent. In addition, agitation may be needed as many bacteria grow better in vitro when agitated, probably to remove from the region of the cell any toxic waste products and bring fresh nutrients to the cell. The composition of the culture medium is also very important. The medium should be rich enough to promote rapid growth of the types of cultures expected to be present in specimen, but care must be taken to balance the nutrients so that the formation of deleterious end-products are minimized. In some instances a special gaseous atmosphere is beneficial. The instrument is required to supply these factors when needed to implement the method chosen. In addition, selection of materials used to manipulate or contain the growing cultures to be free of toxic materials is important.

2. Data Measurements for Each Test Phase

The test procedures imposed on the culture are designed to bring about a measurable change in the cultural growth, cell characteristics, or metabolism when a reaction occurs. It is necessary to know enough about the test being performed, i.e., detection, identification, or antibiotic susceptibility testing to predict what changes may be expected and, if possible, some idea of the magnitude of the change. These changes are then measured as a basis for interpretation and reporting.

a. Detection/Enumeration

Detection of the presence of bacteria in the specimen of body fluids depends on an increase in some measurable parameter, such as the number of particles or the accumulation of some metabolic product during an incubation period. This increase is associated with viable bacteria, i.e., detection of the presence of living bacteria in the specimen. In some approaches semiquantitation is based on the rate of appearance of these measured increases. Depending on the type of body fluid being analyzed, the meaning of the detection of bacteria varies. In a normally sterile body fluid, such as blood, spinal fluid, or aseptically aspirated urine, the presence of any bacteria is significant. In specimens such as clean-catch urine some bacteria are normally present as the result of contamination during collection, and an appropriate number of bacteria (usually $10^5/m\ell$) is considered to indicate bacteriuria. This then requires a time-related response, i.e., the length of time needed for 10^5 bacteria to trigger a positive response by the analytical system used. Other clinical specimens such as wound aspirates, throat cultures, and feces normally have a rich bacterial flora, and would not be suitable test subjects for unselective detection procedures. Selective inhibition might make these specimens amenable to yielding meaningful detection data.

b. Antibiotic Susceptibility Testing

In antibiotic susceptibility tests the growing culture is exposed to a battery of antibiotics or several dilutions of the same antibiotic (MIC) and compared to a control. The antibiotic to which the organism is susceptible may inhibit the growth of the bacterial population either partially or completely while a resistant culture will grow as well as the control. Also, certain antibiotics may alter the morphology of sensitive bacteria. The population growth or inhibition of the culture must be compared to a control and measured by whatever approach is selected — such as light scattering, impedance, etc. In addition, certain methods such as pulse height analysis and differential light scattering, may indicate changes in morphology of the bacterium inducd by certain antibiotics, which would allow an additional measurement for interpretation.

c. Identification

Certain approaches may be used for identification of the predominating bacteria. If the selective inhibitor approach is used, the measurement of growth (or inhibition) in selected media is applied, using methods such as detecting particles by light scattering, or changes in impedance. If serological agglutination is employed, as in the pulse height analyzer, then the evidence of clumping is indicated by increased appearance of large particles as well as by the lack of or slowing down in proliferation of total particles.

d. Types of Tests Performed

The basic pattern in microbiological analyses of body fluid is well established and includes detection/enumeration of pathogens to indicate the presence of an infection and antibiotic susceptibility testing and identification of the pathogen as a guide for treatment. However, the demands for certain types of data change, and at present there is a trend toward performing more MICs and organism identification than in the past. Keeping this in mind, instrument systems that are capable of performing these various types of tests may be more useful in the future. In general, the more flexible the instrument system, the more capable it will be of meeting future trends in microbiological testing.

At present some instrumented systems perform detection/enumeration only, usually specializing in body fluids with a simple microflora, such as urine or blood and spinal fluids. Other instrumental systems are used at present mainly for antibiotic susceptibility tests requiring a pure culture. The ultimate goal, of course, is an instrumented system that can perform a complete microbiological analysis on any body fluid regardless of the complexity of the microfloral population.

3. Control of Biological Processes

The methods proposed for some approaches may include procedures for controlling certain biological processes. For example, if the procedure requires a culture to reach a certain turbidity before it is used as an inoculum, the density of the growth of the culture in the instrument may be monitored, and a mechanism triggered to perform the inoculation when the desired turbidity is measured. As certain microbiological analyses become nearer to complete automation, and need to perform a further step, methods to instruct the instrument to proceed with the analysis must be devised. For example, if a bacteriuria has been detected and the procedure calls for this positive culture to be tested automatically for antibiotic susceptibility, it will be necessary to instruct the instrument that the culture is positive, that it should be diluted × number of times, placed in contact with the proper antibiotics, incubated, and read at the proper time interval. This involves several decisions which direct the mechanical procedures to carry out this test. Thus, process control and decision making become important as more automated processes are included. It is important that all of these methods used for growth, testing, and control of the analysis of microorganisms must be compatible with the capabilities of the instrumented system.

The ideal test methods will employ the appropriate environment for rapid growth, apply an approach capable of performing a complete microbiological analysis on any type of body fluid, and will adequately control the analysis.

C. Data Measurement, Manipulation, and Interpretation

1. Sensitivity

Each test approach generates data in accordance with the type of measurement for which it is designed, such as detecting radioactivity, reading light transmission or scattering, determining electrical impedance, or counting and sizing particles in the test sample; but each type of measurement is useful only if it has the sensitivity, reliability, and reproducibility required to yield meaningful results. It is important that the signal-to-noise ratio yields results that are clear-cut, rather than marginal, and that the differences occur in a short enough time interval to be suitable for a rapid analytical procedure.

2. Measurement Interval

As the test conditions imposed cause changing situations to occur as the culture grows, these must be measured and the time intervals between readings should be spaced according to the expected appearance of useful data. For example, changes taking an hour to occur dictate a reading at that time while if changes are taking place in minutes, several readings made in a short time span are desirable. These data are then collected, reduced, translated, and transferred in a form useful for analysis and interpretation.

3. Criteria

Each approach has its own capabilities for measurement, and among the measurements possible with each instrument the most significant must be chosen as criteria to indicate a reaction. Since a growing culture is used in most approaches, the effect of the test condition will most likely be evaluated at several time intervals by a change in the growth pattern, cell characteristics, or metabolism of the culture when the test condition is applied. These meaningful changes will be defined, as quantitatively as possible and will form the basis for the criteria by which the data are interpreted. As was pointed out in the Microbiological Rules section (Chapter 2), more than one parameter should be chosen. It is important that the selection be based on biological significance of the data and on the sensitivity of the instrumented system to measure the desired parameters. The raw data selected as showing meaningful criteria of a reaction in many instances will need manipulation to present the information in the clearest and most usable form for interpretation.

4. Reports

The data from the selected parameters or criteria compared in an orderly fashion to the control data will serve as a basis of interpretation of the test results. After this interpretation, the information can be assembled into a report. Laboratories in general have a "microbiological jargon" in which the results of the microbiological analysis are reported and the doctor is accustomed to this language. The simplest form in which the results can be presented is usually the best and if the reports compiled in an instrumented system reduce the data to the customary terminology and format, the doctor will then have less difficulty in basing his diagnosis and treatment on this familiar report format.

5. False Results

An important aspect of the interpretation of the data is the relative importance of false negative or false positive results in the detection of the bacterial infection. Most doctors would prefer to have a false positive result which will alert them to a possible infection than to have a false negative result, which may cause a missed diagnosis. A false positive result, on repetition, will probably become negative, and will cost only an extra test. On the other hand, a false negative result, causing a missed diagnosis will certainly delay the proper treatment of an infection, perhaps until it is too late. Also, a false resistant antibiotic susceptibility test is preferable to a false sensitivity, which may result in an ineffective antibiotic being administered, although, when false resistance is reported on an organism resistant to all other antibiotics, treatment will be adversely affected. Also, in antibiotic susceptibility testing, errors compared to traditional methods may be graded as major errors (sensitive vs. resistant) or minor errors (intermediate vs. either sensitive or resistant). Therefore, in evaluating the performance of an instrument, the direction and magnitude of the error is important to consider, as well as the frequency of false results.

Because of the large volume of data generated by the dynamic measurements of many instrumented systems, the automation of data measurement, reporting, and interpretation is a great value. This will be discussed later in this chapter in the section on automation.

In summary, in the collection and management of data from the dynamic reaction, it is important that the signal-to-noise ratio and the data measurement intervals be geared to yield clear-cut results which may be enhanced by data manipulation, interpreted, and reported in traditional phraseology. The interpretation of the data must include an assessment of the significance of false negative vs. false positive results.

III. ENGINEERING CONSIDERATIONS

In designing and evaluating an instrument for microbiological analysis of body fluids, it is important to consider not only the biological requirements but, also, the practicial engineering aspects of the chosen approach. This section is not intended as a "how-to" guide for engineers, but merely an attempt to touch on the important engineering considerations in designing an instrumented system for microbiological analysis, including functions of the instrumented system, special considerations on automation, performance characteristics, configuration, and operational features and limitations.

A. Functions of Instrument

In performing any one or all of the major phases in a microbiological analysis of body fluids, the system must be multifunctional. Each approach will implement the basic functions in a different way, but many of the analytical systems will contain different components to perform similar functions. Automation of the instrumented systems is an important consideration; and the more nearly automated the system, the more functions it will perform.

The functions performed on a specimen for a complete microbiological analysis include:

1. Sample preparation (dilution or concentration)
2. Inoculation
3. Incubation; agitation
4. Measurements or readings
5. Data reduction and interpretation
6. Decision on test conditions to be imposed
7. Sample transfer
8. Addition of test substances
9. Repetition of 3, 4, and 5 above
10. Transcription of data
11. Sample disposal

In addition, the system should include mechanisms for calibration, standardization, detection of malfunctions, cleaning, and sterilization.

The engineer's task is to assemble the proper components to perform the individual functions and to ensure that the components work together as a system to perform an acceptable microbiological analysis. In designing the system, the engineer has to give special consideration to the degree of automation desirable for the system.

The ideal instrumented system would be completely automated and would process the specimen from sample preparation to printing out the report and safely disposing of the samples. Although such a system is not now in existence, some are approaching this goal.

B. Special Considerations in Automation

1. Degree of Automation

The theme of this monograph is a discussion of instrumented methods for microbiological analyses of body fluids with emphasis placed on the more sophisticated decision-making instrumented systems. It has been pointed out that no existing system is completely automated, but that the more sophisticated a system becomes, the more automated processes are included. Thus, a major decision faced by the design engineer and the evaluating microbiologist is, "How much automation is optimum for this particular microbiological assay?"

Keeping in mind the stated goals to be achieved by an instrumented system to perform microbiological analysis of body fluids — *rapid, accurate analysis at a reasonable cost* — the contributions of automation to these goals for each step of the analysis must be considered.

The degree of automation achieved in any instrument system may well rest on the feasibility of interaction between the chosen components. It may equally well depend on the needs of the potential user who should assess which of the functions are required for the intended analyses and determine whether the instrument supplies (or oversupplies) these needed functions. These evaluations will probably determine the degree of automation which best fits the needs of the laboratory.

In order to decide on the degree of automation appropriate for each instrument system, consideration must be given to the necessity and difficulty of automating each of the functions listed above, i.e., what is involved in automating each function of an instrumented microbiological system? An attempt will be made to discuss this aspect, ranking the functions in order of their importance in rapid microbiological analysis, as well as the problems related to their automation.

a. Automation of Measurement or Reading

The very essence of the automation of a rapid microbiological assay is the automatic measurement or reading of the test specimens and the control(s). Rapid microbiological procedures are dynamic in nature, which often entails multiple measurements on each sample; and most instrumented analyses in a routine clinical laboratory generate a large number of readings on many different samples. Unless these instrumented readings are automated, a major value of instrumentation — speed — is lost.

Another important value of automated measurements is the improvement in accuracy by reducing subjective errors, as well as errors in calculation. The sensitivity and precision of automated measurements also enhances the value and accuracy of the results.

The implementation of these measurements, which profoundly influences the automation, is different for the various approaches and will be discussed in connection with each method. However, all approaches have certain features influencing the reading of the samples. This is especially true of the carrier or cuvette for containing the test cultures and controls, which will be discussed later in this chapter under configuration.

b. Data Reduction and Interpretation

Once the measurements have been made by an automated process, the reduction and interpretation of the data by automation is virtually a requirement. This usually is accomplished by a combination of a computer programmed to process the data and a strip chart or X-Y recorder. Under some circumstances, an oscilloscope is needed to display the data for human operators to make an immediate decision.

c. Decision Making

In general, instrument systems that use computers for data processing and interpretation can be programmed for decision making. The necessity for this function is dependent upon the complexity and multiplicity of the analytical processes to be performed. Automated decision making is essential for continuous processing of an analysis consisting of several different procedures or which depend on a choice of reagents such as in antibiotic susceptibility testing. Also, it is important for the engineer to design the instrument to generate data in a form that can be easily translated into simple yes/no for decision making.

To summarize, these three functions — measurement of data, data processing, and interpretation and decision making — are the most important functions benefiting from automation, while other functions are less critical.

d. Incubation and Agitation

In general, microbiological analytical procedures require incubation and many include agitation. Frequently, these two functions are combined in the same unit, and it is highly desirable to include these functions in the automated system. When the reader is contained in the incubator, the sample can be read without interrupting incubation, which is especially important when multiple readings are made on the same sample, or when the measurement is influenced by the temperature of the sample.

e. Sample Transfer

In many analyses, especially multifunctional analyses, it is necessary to transfer the test culture to another carrier for further testing. Thus, the test culture carrier or cuvette may have to be designed to allow aseptic transfer of the contained culture. This places a severe burden on the designer of the culture carrier. However, for multifunctional analytical procedures, the automation of such a transfer function would be most desirable, as it would avoid the necessity of intervention by a microbiologist in a continuing procedure. There must be a careful balancing of the desirability of the automated transfer vs. the difficulty of automating the transfer mechanism within the constraints of the culture carrier. In addition, a provision for making the decision to transfer would be required.

f. Addition of Test Substances

The same factors of desirability and difficulty in the automation of the sample transfer function apply to the addition of test substances.

g. Inoculation

The automation of the inoculation is really desirable as it adds one more capability to the instrument system and relieves the technologist of one more chore, but as the inoculation takes place near the beginning of the test, it may be more convenient to have a technologist available to perform this step manually than in the case of the transfer of the sample which will occur further into the procedure, possibly, after the skilled technologist is off duty. The difficulty of automating inoculation may outweigh the benefits, but problems in automating inoculation may be simpler than those discussed in the transfer procedure.

h. Sample Preparation (Dilution or Concentration)

Sample preparation is the first procedure in a microbiological analysis. Should the sample be prepared automatically, or should preparation precede the start of the instrument involvement? For example, in the detection of bacteria in a body fluid, usually the only preparation needed is a simple dilution (urine) or possibly a concentration (blood). Since the specimens, especially urine, may be received in a variety of containers (urine collecting cups, syringes, tubes, pickle jars), or throat cultures on tongue blades, swabs, etc., a simple manual preparation preceding placing it in the instrument may well be preferable to trying to automate that process. The desirability of automating the sample preparation may be overridden by the difficulty of accommodating the various types of specimens as well as the diversity of collection methods and containers. A true fully automated system for clinical use would have to accept such diverse collecting vehicles.

i. Sample Disposal

At the end of the analysis, the sample must be disposed of. Since the sample may contain living, potentially pathogenic bacteria, disposal entails sterilization. Probably the most expedient way to dispose of the large volume of samples generated is to dump the sample carriers into a container which can be closed and transported to an autoclave for sterilization. Because part of this procedure (transport and sterilization) requires manual intervention, probably only the dumping of the samples can be done automatically. Whether or not this task should be automated will depend largely on its ease of automation.

The many practical problems involved in automating the various functions of a microbiological analysis explain why most of the systems to be discussed are only partially automated. However, the number of automated processes included in the instrumented system are key to attaining the maximum value of instrumentation.

2. Potential Expansion

Among the practical considerations in designing or purchasing an automated or partially automated instrumented system for microbiological analysis is the potential future uses of the instrumented system, with or without additions or modifications. If the principle of action applied in any system is capable of expansion to another procedure in the microbiological analysis, this possibility is worth considering and should be planned for in the initial automation. The engineer must determine what this addition would mean in terms of redesign. Can it be added on in modular form or must the basic design of the instrument be altered? The customer must assess present and future needs. Will the additional capability be needed in the future? Will it mean just adding a module or obsoleting the instrument and buying an expanded version? What will be the cost of this added capability? The customer should be wary of promised future developments, as even with demonstrated feasibility of the additional application, the approval by FDA is not automatic and a considerable delay may occur before the necessary clearance may be granted

C. Performance Characteristics

There are several important performance characteristics needed for instrumented microbiological analysis including speed or test rate, accuracy, reliability, and reproducibility.

1. Speed or Test Rate

Since the name of the game is speeding up the microbiological assay, the approach chosen, and the implementation of that approach should emphasize the rapidity of the method to obtain results by selecting the speediest procedure compatible with proper analysis. The customer should evaluate the need to speed up that particular analysis or portion of the analysis performed by the instrument. For example, a large volume laboratory with many samples coming from out-patient clinics might benefit from having the results of a screentest for bacteriuria before their patients "disappear," while a laboratory with a lower volume of specimens obtained primarily from chronic inpatients might not derive as much benefit from a speedy diagnosis of bacteriuria. Both laboratories, however, might find a bacteriuria test followed immediately by a rapid antibiotic susceptibility test to be valuable in their laboratory.

The speed or test rate of performing the analysis has two aspects. One is the time it takes from the receipt of the specimen to reporting the results. This is of prime importance to the microbiologist and the physician, who wishes to institute treatment as soon as possible. The other factor is the amount of time the specimen is tying up an important functional part of the instrument system. For example, how much reading time is required for each sample? How much incubation time? What volume does the sample have? All of these factors affect the test rate and, in turn, the capacity of the instrument to perform tests and produce results. For instance, the read head may be the limiting factor which regulates the flow (and number) of samples that can be processed in a given time. Could fewer readings or shorter reading times or an additional read head increase the system's capacity? These are engineering choices, but this aspect also becomes the concern of the microbiologist if too small a part of his workload can be processed in a workday. Test rate, therefore, is a critical consideration.

2. Accuracy

It goes without saying that a speedy, but inaccurate instrument, is doomed to failure. The engineer should design into the instrument the controls necessary to ensure accurate results, such as calibration devices, standards, etc. The potential user must assess the accuracy of results produced when the instrument performs the type of assay(s) in which the customer is interested. Biological assays are often inaccurate under the best circumstances. In considering an instrumented system, the question might be, "Would that degree of accuracy be acceptable and improve the results in our laboratory?" The probability of the instrument having reasonable accuracy has improved recently, since the FDA has promulgated standards for performance of certain types of diagnostic instruments before approval for release on the market; but the potential user should still carefully evaluate the instrumental system for acceptable accuracy.

3. Reliability

Engineers are thoroughly aware of the need for the reliable performance of the instrument, and in designing the instrument, components, and systems which are known to be reliable should be chosen. New approaches should be thoroughly tested before incorporation.

The potential user has a special need to emphasize reliability in the evaluation of the instrument. In a clinical setting, tests are requested at all hours of the day or night; and the malfunction of the instrument during a period when maintenance or repair service is uncertain or unavailable can cause a serious delay in obtaining the needed results. The potential user should thoroughly examine the potential for breakdown, the service offered by the seller, and the back-up microbiological material necessary to keep on hand to continue performing assays, if the instrument is inoperable. In smaller or isolated laboratories where an instrumentation repair man is not available, reliability may be the deciding factor in acquiring an instrument system.

4. Reproducibility

The instrument must yield reproducible results ensuring that replicates of the same sample will give the same results. Engineers are well acquainted with the importance of reproducibility of the components of the system and of the whole system; but when doing an analysis to test reproducibility on a growing culture, the instrument must be able either to take duplicate readings simultaneously or apply a correction factor to allow for growth during the time lapse between readings. The microbiologist should consider reproducibility in the same light as accuracy, i.e., can the instrument give better reproducibility than the present procedure employed in the laboratory?

It is clear a fast, unreliable instrument or a reliable, inaccurate system cannot fulfill the goals of an instrumented system for microbiological analysis. A careful balance must be found after weighing all the factors so that an instrument can be designed; and under practical conditions, will perform, microbiological analysis with an optimum combination of speed, accuracy, reliability, and reproducibility. Tradeoffs must be made to obtain the optimum mix of desirable characteristics.

D. Configuration

1. Sample Load

The engineer will be concerned with the problem of fitting together the components needed to perform the desired functions required for each type of approach into an instrumented system of reasonable size, weight, shape, and power requirements. Many factors will influence these requirements — the most important involving the number and volume of the samples to be processed in the system at any one time and the complexity of the processing. The basic data for this figure will be generated by the microbiologist, and will depend on the number of specimens to be processed, the volume of each sample, the timing and number of measurements and the number, complexity, and degree of automated processes.

2. Cuvette

The configuration of one of the components of the system — the test culture carrier or cuvette — is of prime importance in the instrument system design and is influenced greatly by sample size and number, as well as the type of process performed by or in the cuvette.

The culture container or cuvette, of course, must be able to contain the volume of culture required and prevent leakage or spillage; but there are other factors influencing the design, particularly when the culture may be the subject of further tests following detection. For example, a positive culture may be detected in an instrumented system capable of doing antibiotic testing and identification on the same culture. The culture carrier or cuvette must be able either to expel fluid for aliquoting or transfer, or to take up fluids for dilution, or reagents for testing, without leakage or spillage. In addition, the prevention of contamination of the sample and the operator will also influence the configuration and closure of the cuvette.

The container must also be made of sterilizable and disposable material which is nontoxic to the bacterial culture and can be used for prolonged storage of media without contamination. For some approaches, the cuvette must be optically clear, while for others, it must be particle-low. The size of the cuvette will depend on the number and size of the samples coupled with the number of readings.

These requirements for the cuvette must be met while conforming to reasonable size and weight limitations and meeting the configuration requirements of the overall system. These aspects are key to the automation of the microbiological process.

3. Size, Weight, and Power

When the detailed test procedure has been finalized and the number of samples estimated, the engineer can calculate the size, weight, and power requirements needed to do the desired workload. If these calculations lead to an unreasonable configuration, tradeoffs must be made to obtain the best configuration, possibly leading to a modular system. Additional considerations are the type and placement of control knobs on panels to meet human convenience and inclusion of alarm signals.

The potential user is concerned about laboratory space and power requirements, accessibility, and convenience in use of operational features.

E. Operational Features and Limitations

1. Contamination by Infectious Agents

Design of instruments for microbiological analysis of potentially infectious specimens poses special problems for the engineer. The design must prevent contamination of operator by the culture, including aerosols of the culture. Also, safeguards must be incorporated which prevent the operator from gaining access to the interior of the instrument, in the event of an accidental contamination, until the area has been completely decontaminated. The instrument design must also prevent contamination of the specimen by the environment, as well as include a mechanism for destroying and cleaning up of the potentially infectious material in the event of a malfunction of the instrument or a spill of the culture. A precautionary decontamination of the instrument after use should be considered.

The microbiologist must carefully assess the adequacy of the precautions to prevent or control contamination. An understanding of just what decontamination procedures are appropriate should be established and evaluated for feasibility under the prevailing laboratory conditions.

2. Overall Safety

In addition to microbial contamination, there are other safety factors to be considered, of which the engineer should be aware. For example, radioactive material may be used for certain approaches (radiometric assay) and proper containment procedures and provision for radioactive waste disposal must be considered.

The microbiologist must carefully assess any safety hazards and provide for handling of any accidents. In the case of radioactive reagent use, the pathologist should ascertain rules and regulations regarding the shipping, receipt, handling, storage, and disposal of radioactive material and determine whether the laboratory can handle these requirements.

3. Ease of Operation

The engineer must consider the usual means of designing a smoothly operating instrument system which allows the user maximum convenience and efficiency, including a simple control panel with the minimum number of control knobs and clearly designated warning signals to indicate malfunction.

The chief microbiologist should consider the caliber of the potential users, including night and weekend personnel, and evaluate the training necessary for this personnel to operate the instrument correctly. He should ask for a training guide, and pay special attention to the design of the control panel and the "confusion factor" presented by panel and display with warning signals that are easily interpreted.

4. Maintenance and Serviceability

The engineers should, as usual, strive for an instrument system that is as maintenance-free as possible. Those maintenance items which can be accomplished by the user should be convenient and user-oriented, including easy access to and replacement of the item. The items to be maintained by service personnel should also be designed for convenience, with only servicemen that know and understand the equipment involved in servicing and maintaining it.

The microbiologist should be aware of the types of maintenance and service required by the instrument and determine what services are provided and how promptly responses for requests for service may be expected. In addition, the microbiologist is responsible for evaluating the help which may be at hand, such as laboratory service personnel or local servicemen, and have the instrument inspected to determine ease of serviceability and the practicality of maintaining a stock of components most likely to need replacement. The microbiologist should then consider the whole picture with respect to maintaining and servicing the system while evaluating its utility in his particular laboratory.

IV. MONETARY CONSIDERATIONS

Monetary considerations are important to the engineer planning the design of an instrument system, to the manufacturer of the system, and to the potential user. Obviously, all are businessmen hoping to profit from their efforts. There are several factors involved in the cost considerations, including the original cost of the instrument and consumables to the manufacturer and to the consumer, as well as the overall cost per test which is particularly important to the potential user.

A. Cost of the Instrument

The overall cost of designing, developing, manufacturing, and marketing the instrumented system balanced against the probable return on investment is crucial in the decision of any company to enter the market with such a system. The engineer plays a major role in the cost analysis of this project and in producing an instrument at a reasonable price by being cost conscious in the components used to implement the desired functions. The major costs of the system to the manufacturer are those connected with the research, development, production, and marketing of the instrumentation. The latter two are much more easily calculated than the research and development.

In evaluating the cost of the system the microbiologist must look comparatively at competing instruments, as this will represent a sizeable capital expenditure for the laboratory, but will be incurred only once. Lease/rent possibilities may be explored, as the first generation of instruments is often quickly obsoleted by improvements and a lease or rental arrangement allows for an orderly acquisition of the second generation system. Also, the types and scope of service contracts offered should be considered when evaluating the instrument cost.

B. Cost of Consumables

The engineer has relatively less to consider with respect to the cost of consumables, although the design of a novel and reasonably priced cuvette or media carrier may well contribute substantially to the uniqueness of the system. Keeping the cost of manufacturing the consumables as low as possible is important, as these items are the repeat money makers for the producer of the system.

The potential user should place emphasis on the cost of the consumables, as these items will constitute a continuing expenditure, and in evaluating the shelflife vs. probable use to determine in what quantities the consumables will be used; and, therefore, what possible discounts can be anticipated. The reliability and speed of delivery of consumables in case of emergency should be explored as well as the amount of quality control required of the consumer.

C. Overall Cost/Test

The system will probably be evaluated by the potential user most critically on the basis of the cost/test compared to the cost/test using the conventional procedure. A true evaluation requires the consideration of several diverse factors. Of course, the cost of the instrument plus service added to the cost of the consumables used for each test accounts for a sizeable part of the cost/test.

However, another big factor is the cost of the labor required for each instrument-performed test vs. the classical method. In addition, the reduction in working, incubator, and storage space realized when an instrumented analysis is used should be considered. It is also not unrealistic to credit the more rapid instrumented procedure with increasing the number of specimens received for analysis which would spread the cost of amortizing the instrument and overhead over more tests.

The sum of all these monetary considerations may yield a surprising result.

It is clear that the production of the instrumented system and consumables are of primary importance to the seller, whereas, the cost/test compared to the traditional tests is of greatest interest to the consumer.

In summary, perhaps the most important, practical consumer consideration of all is how well the instrument system meets the present overall needs of the laboratory. Considered in that evaluation are the volume of work, the type of tests to be performed (which can range from a rapid screen test for bacteriuria to a complete detailed analysis), the cost per test compared to the traditional system, the maintenance and service of the instrument, and the caliber and training of personnel required to operate the instrument. These factors will define the class of instrumented system to consider. Further evaluation is based on the ability of the instrument to meet the theoretical and practical goals of instrumented microbiological analyses and the additional future needs of your laboratory must be matched with the potential of the instrument to fulfill these needs.

In addition to the overall practical aspects which apply in general to instrumented microbiological analytical systems, there are certain practical considerations that are of special relevance to instruments embodying the different principles of action. These will be discussed in the chapters relating to each approach.

V. PRACTICAL ASPECTS OF EDUCATION RELATED TO INSTRUMENTED MICROBIOLOGY

One of the very real practical considerations with respect to instrumented systems for microbiological analysis is how to educate the various persons who need basic information in every step of the process, from decision to produce, to purchase, and use of such a system. This includes such diverse groups as management and technical personnel in the commercial firm considering production of the instrumented system, the clinical laboratory scientists evaluating and using the system, and the physicians receiving diagnostic and treatment guidance from the reports generated by the system, as well as the general public.

A. Personnel of Commercial Company

The decision to embark on a program for research development and production of an instrumented system for clinical microbiological analysis will be made at a high level of management because of the long-term, high-cost efforts involved and will be based on the opinions of a group of senior advisers of various disciplines, each needing basic information on instrumented systems for microbiology.

Marketing plays a key role. The need for such instrumented systems has presumably already been established but volume of sales must be forecast both of the instrument and the disposable items, as well as the probable life of the product. Another marketing question is how well the instrument system fits into the present product line. The financial aspects are very important and will include the total cost and time span before return on investment is realized, as this impacts the total budget and cash flow. In addition to the traditional costs, a new monetary consideration will be the expenses connected with qualifying the instrumented system with FDA. There are other questions which must be answered. Are the total resources available to launch this program? What are the opinions of the medical advisors on desirability, the engineers on feasibility, and the microbiologists on the practicality of the proposed system? The advice of the legal department on the patent situation is valuable, especially if expertise or a prototype instrument may have to be acquired.

After all of these factors have been carefully weighed and a decision is made to go forward, the project manager must form a team to perform the work. Frequently, there are objections to this project from the team or their advisors for various reasons. The R & D microbiologists are usually too far removed from the clinical laboratory to appreciate the advantages of the instrumented microbiology in speeding up results, lightening the workload, and relieving tedious repetitive operations. They see only an interruption of their ongoing project which they understand, in order to introduce a new, unknown program into their work schedule. The marketeers are used to selling conventional products and fear the uphill battle needed to educate and sell this new type of product to their customers. They are also troubled by the difficulties of making a reasonable estimate of the potential market. The medical advisors to the manufacturers assume their usual cautious attitude, which is hard to argue with convincingly, as the fledgeling instrument system cannot as yet prove itself, and the budgetary types usually recoil in horror at the projected, long-term expenditure of large sums of money for a product aimed at a far-distant market.

What can the project leader do to sell the instrumented system to the project team and advisors? The answer is education. First, make available to all members of the team and the advisors the basis on which management decided to initiate the project, followed by a thorough discussion of the advantages of the product to the microbiology laboratory and the doctors. Then set up a plan of action, acquaint the team members with their role, stressing the importance of their contribution, and keep constant check on the program. Advise all team members and advisors of the progress and problems encountered and enlist their help. Finally, inject a belief in the project in all company personnel, because only an enthusiastic group can successfully bring a complicated instrumented product to market.

B. Laboratory Microbiologists

Once the instrumented system is on the market, a new group needs education — the pathologist and/or the chief microbiologist needing information to evaluate the instrumented system and the technologist who will use the system.

The clinical laboratory personnel in the microbiology department is often upset by the prospect of "being replaced by a machine". The chief microbiologist sees problems in introducing a new routine into the laboratory, particularly a method that is such a radical departure from the traditional procedures. The whole concept of instrumented microbiology represents an unknown and places an extra burden of learning, training personnel, and, maybe, even servicing the mysterious "black box". Worry about breakdowns and the performance of night and weekend shifts add to the chief microbiologist's resistance.

What can be done to gain the confidence of the microbiology personnel and further their acceptance of this new method? First, the manufacturer should thoroughly acquaint the pathologist and chief microbiologist with the advantages of the instrument and its operation, and institute a training program for all personnel involved, including teachers of medical technology and supply an adequate number of detailed operating and service manuals. Also, the ready availability of service personnel should be stressed. The chief microbiologist should emphasize the positive side of the new routine and, by pointing out the improvement in obtaining timely results, the relief from tedious chores, as well as by encouraging the technologists to talk to the chemists who were involved in a successful transition to automation to get their reaction. The technologists should be thoroughly trained on the instrument so that the operation is familiar to them. For the instrumented system to succeed, it is necessary to have the cooperation of the whole microbiology department.

C. Doctors

After the laboratory has acquired a new instrumented system, the doctors who use the laboratory need indoctrination. Frequently, a doctor is disinterested in the addition of rapid instrumented procedures in the microbiology laboratories because they do not fully appreciate the benefits. To really make full use of the rapid test results may mean an alteration in their schedule of hospital rounds or an extension of rounds, which may not appeal to them unless they see a benefit. In addition, computerized reports may not be as familiar as the previous reporting method and may be confusing.

To get the doctor's enthusaistic support of the new procedures, an effort must be made by the detailman and the pathologist or microbiology chief to show the doctors how rapidly the results can be available and how his patient will benefit by earlier treatment. Also, point out that complete testing with immediate application of results may ward off malpractice suits. Attempts to make the doctor aware of this new service might include seminars, demonstrations, or personal invitations to visit the laboratory. The reports should be explained fully to familiarize the doctor with computer printouts, and any new information generated should be pointed out. A large part of the successful use of rapid instrumented methods depends on the doctor's attitude and use of the improved service, so he must become involved in a positive way.

D. Education of the Public

The education of the public in medical subjects may be controversial, but the informed patient is usually more cooperative, and the realization that steps are being made to speed up his treatment by instrumented systems may improve the doctor-patient relationship.

The education of the public in advances in health care may be handled by qualified professionals through such programs as, "The Public Needs to Know," which is a regular feature of the American Society for Microbiology's annual meetings, and by inviting laymen to select presentations offered at local medical association meetings. At such gatherings, questions could be answered by professionals giving authoritative information to satisfy the public's need to know about medical progress.

It is evident that the educational program is important for all persons concerned with the production or use of systems for microbiological analysis.

VI. PRACTICAL ASPECTS OF EXTENDED OPPORTUNITIES IN MICROBIOLOGY

Several practical aspects not previously considered center around the opportunities that instrumented microbiological systems offer for the extension of microbiological service and knowledge by yielding results not attainable with present procedures.

A. Extension of Service in Microbiology

1. Life Saving

In the field of service, the lifesaving potential of the truly rapid instrumented methods is of paramount importance, especially for the increasing number of compromised patients whose normal resistance to infection has been lowered by therapy with immunosuppressive drugs. Certain rapid instrumented methods yield antibiotic susceptibility results to guide treatment within a few hours after sample collection with a rapidity which is unattainable by present methods. This greatly extends the doctor's capability to save the patient's life.

2. Malpractice

Rapid instrumented systems also extend service by giving accurate test results which can act as a safeguard, if necessary, in malpractice suits.

3. Humanitarian

At the risk of appearing "square", the benefit of rapid, accurate, instrumented test results in alleviating human suffering should be pointed out. Even relatively minor infections when not treated promptly and properly, can mean discomfort for the patient and possible loss of work and wages. For families with a marginal income, the loss of even a day's pay can result in some privation, often with social impact. The adequate treatment of even minor infections could cut down on human suffering and be of benefit to mankind.

B. Extension of Knowledge

Instrumented systems of microbiological analysis offer advantages in studying microbiological reactions, because large amounts of data can be generated and analyzed quickly and easily, thus expediting certain types of research and extending microbiological knowledge. But, more importantly, certain instruments may make possible experiments on microorganisms that cannot be performed by existing procedures. The possibility of exploring new frontiers in microbiology will aid in interpreting data from microbiological analysis of body fluids and in perfecting new tests to diagnose other infectious diseases. This aspect will be discussed more thoroughly in a later section.

REFERENCES

1. **Bailey, W. R. and Scott, E. G.,** Determination of susceptibility of bacteria to antimicrobial agents, in *Diagnostic Microbiology,* C. V. Mosby, St. Louis, 1970, 294.
2. **Hobson, P. N.,** Growth of mixed cultures and their biological control, in *19th Symp. Soc. Gen. Microbiol.,* Meadow, P. and Pirt, S. J., Eds., Cambridge University Press, 1969, 43.
3. **Stanier, R. Y., Adelberg, E. A., and Ingraham, J.,** *The Microbial World,* 4th ed., Prentice-Hall, Englewood Cliffs, N. J., 1976, chap. 9.

Chapter 4

EARLY INSTRUMENTED SYSTEMS FOR MICROBIOLOGICAL ANALYSES OF BODY FLUIDS

I. INTRODUCTION

Many clinical pathologists and microbiologists will be called upon to decide whether to acquire a certain instrument or system available for microbiological analysis. What guidance do they need to make an intelligent decision? First, they need to understand the principle of action of the instrument and to evaluate how well this instrument fulfills its intended analytical function. The theoretical and practical considerations behind the design of instruments for microbiological analysis presented in the previous sections will aid the microbiologist-pathologist in understanding the requirements for an instrument system to perform these analyses. The following sections will discuss several approaches used in various sophisticated instrumented systems intended to perform one or more of the major phases of a microbiological analysis and can form the basis for the evaluation of acceptability of the instrument.

This discussion is not intended to be an exhaustive review of all the sophisticated instruments and systems used in microbiological analysis but will attempt to select instruments illustrative of the various approaches, both historical and presently active. Some of the instruments included in this discussion were developed, marketed and withdrawn, while others never reached the market due to apparent shortcomings, either technical or economic. Some instruments or systems are being marketed successfully, while others are waiting in the wings for completion of development or approval for release. An attempt will be made to discuss the reasons for success or failure of those that reached the market and the status of those nearing completion.

II. HISTORICAL PERSPECTIVE

An awareness of the need for instrumentation in microbiological analysis became evident in the mid- to late-1960s and has been increasing ever since. A government sponsored conference was held in April 1966 on the subject of rapid identification of biological agents during which reports were presented on several instruments aimed at one of the major phases of microbiological analysis — bacterial identification.

A. CPE

At this meeting, Strickler and Gafford[1] of Beckman Instruments described an instrument for rapid bacterial identification based on the principle of continuous particle electrophoresis. A curtain of electrolyte containing the sample flowed downward past electrodes causing separation of bacterial particles on the basis of surface charge into specific identifiable bands in the spectrum according to bacterial type. This instrument (CPE) was marketed but did not gain wide acceptance for bacterial identification, probably because the spectra of certain bacteria overlapped making a clear-cut identification difficult. No results of clinical trials were published.

B. BAIT and TAMIS

Another paper by Davis and Glenn[2] reported on two automated systems, BAIT and TAMIS, developed by the U.S. Air Force for field use in detection and antibiotic susceptibility testing for biological warfare agents and, subsequently, for diagnosis of infectious disease under primitive combat conditions. The detection and identification tests depended on the collection of the infectious agent and extraction and reaction of the antigen with specific antisera in a narrow tube filled with a semisolid medium (Ouidin column). The precipitin reaction produced was scanned and the pattern was compared by telemetry with a bank of known patterns (TELUS)[3] at a central Air Force base to identify the organism. This novel system worked well for its intended use, but had limited appeal, being mainly for the military, and never appeared on the commercial market. It was not aimed at a normal clinical laboratory, but was designed for microbiological monitoring under combat situations.

C. Colony Scanner

At this same time, Glaser[4] published on his mechanized, computer-assisted colony counting and identification system, based on an optical scan of the colonies on an agar surface. Bacterial identification was based on the colony characteristics.

A recent version of the instrument system[5] is two stories high and automatically passes the agar-filled trays with the developing colonies in front of cameras for photographing. By computer analysis of the photographs, colonies are selected, picked, and inoculated for further study. These colonies can also be visually studied through windows in the instrument, and the temperature, humidity, gaseous environment, and the light in the sterile incubator can be varied for experimental purposes. Also, a variety of agars can be made in an agar-making plant associated with the system and distributed onto different trays. The growing colonies can be exposed to drugs, such as antibiotics, and susceptibility can be determined in about an hour. This instrument is used primarily as a research tool to detect mutants in genetic studies and this concept has not been translated into a commercial product for clinical laboratories, although it has been proven to have speed and considerable accuracy in identifying pathogens. No clinical trials have been conducted.

D. MAS

With NASA funds (NASA Contract NASW1575, 1967) in IBM laboratories, Gall, Clayton, and Wallace[6] designed and, with IBM funds, built a breadboard model of a partially automated robot microbiological analytical device (MAS) intended to monitor the microbiological status of the astronauts and their environment during space flight when no trained microbiologist was aboard. The instrumented system was intended to prepare, inoculate, and incubate the sample mechanically; count and color-match colonies grown on selective media intended for further testing; and under electronic control mechanically transfer selected colonies to specialized media for identification or antibiotic susceptibility testing. After incubation, results of these tests were read electronically, mainly by color-matching. The instrument was programmed by a sequencer to perform the analysis automatically. MAS was constructed primarily with the needs of the space flight environment in mind and was not developed further for terrestrial use, as it was impractical for a clinical laboratory where trained microbiologists were available.

E. SEROMATIC®

In 1969, Aerojet General described[7] the automated system, SeroMatic®,* which performs all steps needed to prepare slides for the fluorescent treponemal antibody absorption (FTA-ABS).

Patients' serum is placed on slides pretreated with antigen which are fed into the

* ®Registered trademark of Aerojet General Corp., Aerojet Medical and Biological Systems Division, El Monte, Calif.

instrument for mechanical transfer through the incubation, washing, rinsing, and drying cycles and the application of the fluorescent conjugate, after which the slides were read as usual under the microscope. Claims were made of a fivefold increase in the technician's work capacity to 200 slides per day, and the comparison of results with the standard methods was good.[8] Fischer Scientific sold and serviced the instrument for a while but withdrew after disappointing sales. This may have been due to the fact that the instrument was noisy, cumbersome, and did not perform the most critical part of the test — the reading of the slides. Also, the cost per test of the reagent was high, compared to the conventional VDRL test. The rather high price of the instrument was reduced drastically while it was on the market, apparently with the expectation that the sale of disposables was the real moneymaker, but this strategy did not save the instrument from withdrawal from the market. (This is an important factor for a customer to remember. The initial purchase or lease/rental price of the instrument is probably a small part of the cost of doing the tests. The constant replacement of consumables is the real expense.)

III. SYSTEMS REACHING ADVANCED DEVELOPMENT — NOW TERMINATED

More recently, several instruments intended to perform one or more parts of micribiological analysis have been developed to an advanced stage but have never achieved a position in the marketplace. These include the capillary tube scanner, TAAS® (Technicon), DuPont Luminescence Biometer®, and Instrumentation Laboratories microcalorimeter.

A. Capillary Tube Scanner

The fundamental principle of operation of the capillary tube scanner is that of a life detection system predicated upon the detection of new light scatter points within a nutrient-agar filled capillary tube which are attributable to the growth of bacterial colonies. The simple, sensitive, technique allows detection of organisms such as *Escherichia coli* in approximately 2 hr; and, in addition, the instrument can be used for identification and susceptibility testing.[9]

The instrument is essentially composed of a series of glass capillary tubes which are filled with inoculated agar and passed through an intense light beam.[10] Particles in the agar scatter the light which is sensed by a photo-detector so positioned that it receives none of the direct illumination. The scattered light results in electrical pulses which are related to the particle size, allowing the counting and sizing of the particles.

The instrument is capable of becoming a totally automated life detection system and offers the significant advantage of flexibility.[9] A variety of different determinations may be performed, including the detection of organisms, i.e., evaluation of sterility, the enumeration of organisms in clean but nonsterile preparations, bacteriuria screening, and identification of bacteria by selective reagents.

The main thrust of the work was the adaptation of this system to antibiotic susceptibility testing.[11] A series of separate agar preparations of a given isolated organism was made containing a variety of antibiotics at various concentrations, and sensitivity determinations were performed by comparing the growth in the various antibiotic-containing tubes with the growth in control tubes containing no antibiotic. Separate arrays of antibiotics can be used for gram negative and gram positive organisms.

The instrument itself can be designed to accept all of these determinations in any random order simply by placing a machine-readable code on the sample plate.[9] A mechanical system can be provided to automatically remove the plates from an incubation unit, scan them, and return them to the incubator. A small computer could easily and, today, fairly inexpensively, keep track of the various light scatter counts obtained on initial and subsequent incubations for the variety of different types of determinations and produce the desired reports as soon as sufficient information is obtained.

The techniques used with this instrument really depart very little from well accepted conventional methods and should be readily accepted by microbiologists because of the advantages of automation. First, the technique is basically an agar dilution technique, which avoids some of the difficulties associated with the commonly used disk diffusion techniques. Also, since it has the capability of detecting bacterial growth when the colonies are extremely small, there is the theoretical potential of evaluating sensitivities of mixed inoculae.

This basic work was done at the National Institutes of Health in collaboration with Bowman and Vurek in the late 1960s.[10] Bowman holds U.S. Patent 3,574,063 on the basic light scatter system. Later, at the University of Minnesota, under a National Institute of General Medical Sciences grant, a disposable plastic sample plate was developed and patented in U.S. Patent 3,925,166 by Blume and fabricated with assistance from American Instrument Company. This sample plate, essentially, is the geometric equivalent of an array of capillary tubes, each of which can be preloaded with an antibiotic preparation to facilitate the practical use of this concept for sensitivity determinations. Prototypes of the instrument system were made by Consolidated Electrodynamics, Division of Bell & Howell Corporation, and by American Instrument Company.

Development of this instrument was stopped when the government funding was not continued, due to the feeling that private industry should support this type of profit-making development. However, no commercial company has retained a proprietary interest in the system.

B. TAAS®

In 1970 Technicon Instruments Corporation, Tarrytown, N.Y., introduced TAAS*® (Technicon Automated Antibiotic Susceptibility System), which was the first semiautomated instrument system for antibiotic susceptibility testing in clinical microbiology, reducing to 3 hr the time to perform an accurate antibiotic susceptibility test after obtaining a pure culture. The system used a very sensitive diffuse multiangled densitometer system to count the individual bacterial particles from a pure culture of standardized inoculum. The principle of action relating to photodetection is discussed in the section on theory.

The optical system consists of a reverse dark field with Koehler-type illumination constructed with a narrow illuminating angle. This design results in a precisely defined and uniformly illuminated sensing volume with a relatively large depth of focus and very little stray scattered light or "noise".[12]

The inoculum is dispensed automatically and tested with 13 antibiotics in separate wells in disposable plastic trays. After a 3-hr growth period the cultures are killed with formalin, and the growth in the antibiotic-containing tests is compared with the controls by counting the light scattering pulses of individual bacteria passing through an illuminated counting chamber. The results are expressed as a ratio of the population increase of the test sample to the population increase of the control less the zero hour control. The instrument is fully automated after sample introduction and can handle 40 specimens/hr.

Claims for the instrument made in the product literature included speed (savings of a full day), accuracy, reproducibility, reliability, standardization, and reduced cost per

* Registered trademark for Technicon Instruments Corporation, Tarrytown, N.Y.

test. Clinical tests done by Isenberg[12] showed that the test was rapid and that there was good correlation between the results with TAAS® and the Kirby-Bauer procedure. Yet this program has been discontinued, possibly because of the high cost of the instrument to the consumer and large size of the system.

C. Bioluminescence

In about 1970, NASA research on life detection using the bioluminescence approach was applied to the analysis of urine specimens for bacteriuria by workers at Goddard Space Flight Center.[13]

NASA patented the technology and the device to perform bacteriuria tests and several companies produced the instrumented system. DuPont Instruments, Wilmington, Del. 19898, led the field in the promotion of this application using its Luminescence Biometer 760®, an extremely sensitive photometer which detects and measures the light produced during the luciferin-luciferase reaction in the presence of adenosine triphosphate (ATP).

Bioluminescence refers to the production of light as the result of a biochemical reaction produced by organisms, involving an oxidation with enzymes-luciferases of a substrate-luciferin. This reaction requires the energy-rich compound, ATP, which is found almost universally in bacteria in approximately equal amounts per cell. When ATP is liberated from the bacterial cells and added to a luciferin-luciferase mixture, light is produced in proportion to the ATP present, which can be measured in a photometer.[14] Thus, from these data, the biomass, i.e., the number of cells, can be calculated.

In clinical microbiology, bioluminescence techniques have been used to detect and enumerate bacteria in urine[15] and to do antibiotic susceptibility tests.[16] Detection of bacteria depends upon preparing the specimen to eliminate nonbacterial ATP and then liberating and reacting the ATP from the bacterial cells with luciferin-luciferase. The bacterial ATP is then measured so that the presence of bacteria is indicated and the numbers of bacteria can be calculated, based on the average ATP content per bacterial cell.

Antibiotic susceptibility tests are based on the increase in total ATP produced by the growing control and resistant cultures as compared to none or a smaller increase with the susceptible cultures.

The procedure for performing a bacteriuria test on a urine specimen consists of getting rid of the nonbacterial ATP by chemical means, washing and concentrating the bacteria by filtration, and then enzymatically releasing the bacterial ATP, which is measured by light emitted in the Luminescence Biometer 760® as the luciferin-luciferase-ATP reaction takes place in a cuvette within the light-tight instrument. The number of bacteria is calculated based on the average bacterial cell content of ATP and if there are 10^5 bacteria per milliliter urine or more, the specimen is designated as positive for bacteriuria. In a study conducted on 2000 specimens for bacteriuria, no false negatives and only 2% false positive results were attained.[17] The false positive results possibly were caused by the fact that dead bacteria, L-forms, and mycoplasma which contain ATP were present in high enough numbers in the false positive specimens to increase the biomass count, while not forming countable colonies on the plates. Bacteria which form clumps may give a falsely low colony count, as several bacteria clustered together may yield only one colony; but each bacterium in the clump has its usual quota of ATP, which will be released and measured during this test. This may also account for "false" positives as judged by colony counts.

The features of the Luminescence Biometer 760®, as expressed in DuPont product literature, include a rapid test procedure — bacterial count in 10 min, ATP reading in

10 sec, high sensitivity — measures 10^{-13} g ATP or 1000 bacterial cells, and semiautomation after sample preparation.

There are some practical considerations in connection with this test, including the length of time to prepare the sample; the fact that dead as well as living bacteria contain ATP which is measured by the outlined procedure, and the cost of the rather expensive reagents and the manual labor in sample preparation. In addition, each day the luciferin-luciferase reagents must be standardized.

The Luminescence Biometer 760® is no longer being marketed for clinical application for microbiological analysis of body fluids due to recent FDA regulations which caused DuPont to discontinue its marketing effort for this application.

D. Microcalorimetry

Microcalorimetry is a measure of small heat changes and by plotting the heat produced (thermogenesis) against time, a heat profile (thermogram) is created. Microorganisms produce small amounts of heat during growth and metabolism which can be recorded continuously on a strip chart to produce a tracing which is designated as the heat profile.[18] Heat production varies with the stage of growth, starting when the culture reaches log phase, and continuing during growth, until the culture reaches the death phase. The amount of heat produced parallels the metabolic activity, which declines as the culture becomes less active. The changes in metabolic activity, and hence heat production, may be quite abrupt and will be reflected in the heat profile, which is characteristic for some bacteria.

Since microcalorimetry can detect the small changes in heat brought about by metabolism of microorganisms, this measurement can form the basis for determination of growth or inhibition of growth under experimental conditions. Growth is indicated by the production of heat in the culture medium. Antibiotic susceptibility is indicated by lack of heat production in the presence of the antibiotic while resistance is found when the heat produced equals that of the control, according to Binford et al.[19] This method also differentiates between bacteriostatic and bactericidal antibiotics, the former reducing but not eliminating heat production, while the latter stops heat production completely. The most studied application of microcalorimetry is identification of common pathogens, such as Enterobacteriaceae, by means of their different thermograms. Each type of bacterium produces a characteristic heat profile which can be used to identify the unknown bacterium by referring to a library of heat profiles related to known bacterial species.

Some microcalorimeters are based on thermal flux which maintains temperature stability to insure accurate results. Heat flows along a thermopile to a heat sink and is measured as the temperature gradient along the thermopile. The data from these measurements are processed by a computer which constructs the heat profiles, or thermograms. The thermograms are recorded on the strip chart as analog tracings which allow ready comparison and interpretation of the data.

There are several instrumented systems designed to apply microcalorimetry to microbiological analysis. Instrumentation Laboratories, Inc. of Lexington, Mass., has conducted extensive studies on developing an instrument to analyze body fluids in a routine clinical microbiology laboratory.[20] This instrumented system is aimed particularly at identification of Enterobacteriaceae, and consists of an isothermally jacketed, 50-channel device held at 37°C in which the heat produced is measured. Data acquisition, processing, and display and instrument control are computerized.

The analysis is simple, sensitive, and rapid. It involves placing an inoculum of about 500 cells of a diluted, log growth culture in a small amount of a designated liquid medium in a small tube, growing it in the calorimeter, and automatically measuring the heat produced, from which a heat profile is constructed. The characteristic profile serves to identify the organism by comparison with profiles of known bacteria.

The medium used for identification must be very consistent in composition, with even lot-to-lot variations in media produced by the same company giving equivocal results. Since identification depends on comparison of the unknown with a reference library of heat profiles, the standardization of the media composition is of paramount importance.

The method of measuring heat production must be of sufficient sensitivity to detect the small amount of heat produced by microbial growth above the background. Since most bacteria have an output of 40 to 60 μ cal/(sec)(mℓ), backgrounds in the order of 0.8μ cal allow a reliable measurement. Instrumentation Laboratories instrument is very sensitive, as it can detect heat changes from 10^5 to 10^6 organisms per mℓ. Most organisms reach their maximum output of heat and return to the baseline within 5 to 7 hr. Some produce a recognizable profile in 3 hr after start of heat production, while others require up to 14 hr. In a study with this system, 200 cultures of 47 species from 24 genera of clinically significant bacteria were identifiable.[20]

Although this instrument was in the advanced stages of development and had shown reasonably good results, the decision has been made to abandon further work on this instrument system, possibly because of the stringent requirement for media of uniform composition for bacterial identification.

A study was conducted at the University of South Florida, Tampa, Fla., by Binford et al. applying microcalorimetry to antibiotic susceptibility testing.[19] A flow modification of the Beckman Model 190B® Microcalorimeter was used to test eight antibiotics against 15 types of bacteria. Under aerobic conditions, 87% correlation was found with the results of the disk agar diffusion method. These tests also showed that the method distinguished between bacteriostatic and bactericidal drugs.

LBK, Rockville, Md., markets a Microbiological Flow Calorimeter® to study and monitor the growth and metabolism of microorganisms. Their product literature states that it is a fast, sensitive, and efficient analytical instrument with the potential for development and use in a routine clinical microbiological laboratory for identification and enumeration of species and antibiotic susceptibility testing. The present instrument is not suitable for routine clinical microbiological assays as it processes too few specimens per day.

None of the instruments or systems mentioned in this chapter on early systems for microbiological analysis developed before 1970 has made a successful market penetration. However, they do represent several novel approaches, most of which were technologically feasible, and offer valuable experience to the scientists, engineers, and marketeers working on more recent systems.

REFERENCES

1. **Strickler, A. and Gafford, R. D.,** Rapid Identification of Microorganisms by Continuous Particle Electrophoresis in Proc. Conf. on Rapid Identification of Biological Agents, Davis, I. and Williams, R. E., Eds., U.S.A.F. School of Aerospace Medicine, Brooks Air Force Base, Tex., and Office of Naval Research, Washington, D.C., 1967, 100.
2. **Davis, I. and Glenn, W. G.,**Instrumented Identification for Bacterial Differentiation. I. The BAIT and TAMIS Concepts, in Proc. Conf. on Rapid Identification of Biological Agents, Davis, I. and Williams, R. E., Eds., U.S.A.F. School of Aerospace Medicine, Brooks Air Force Base, Tex., and Office of Naval Research, Washington, D.C., 1967, 138.

3. **Glenn, W. G., Russell, W. J., Prather, W., and Jaeger, H. A.,** Instrumented Identification for Bacterial Differentiation. II. Immunoelectronic Principles and Instrumentation, in Proc. Conf. on Rapid Identification of Biological Agents, Davis, I. and Williams, R. E., Eds., U.S.A.F. School of Aerospace Medicine, Brooks Air Force Base, Tex., and Office of Naval Research, Washington, D.C., 1967, 145.
4. **Glaser, D. A. and Wattenburg, W. H.,** An automated system for the growth and analysis of large numbers of bacterial colonies using an environmental chamber and a computer-controlled flying spot scanner, *Ann. N.Y. Acad. Sci.*, 139, 243, 1966.
5. **Glaser, D. A.,** Automation of colony identification and mutant selection in *New Approaches to the Identification of Microorganisms*, Hedén, C. F. and Illéni, T., Eds., John Wiley & Sons, New York, 1975, chap. 1.
6. **Gall, L. S., Clayton, W., and Wallace, C. T.,** Partially automated system for microbiological analysis in *Developments in Industrial Microbiology*, Vol. 11, Corum, C. J., Ed., Garamond/Pridemark Press, Baltimore, Md., 1970, 460.
7. **Mishuck, E. and Roberts, M.,** Automation of fluorescent antibody techniques in *Automation in Microbiology and Immunology*, Hedén, G. C. and Illéni, T., Eds., John Wiley & Sons, New York, 1975, 464.
8. **Lewis, J. S., Duncan, W. P., and Stout, G. W.,** Automated fluorescent treponemal antibody test: instrument and evaluation, *Appl. Microbiol.*, 19, 898, 1970.
9. **Blume, P.,**personal communication.
10. **Bowman, R. L., Blume, P., and Vurek, G. G.,** Capillary tube scanner for mechanized microbiology, *Science*, 158, 78, 1967.
11. **Blume, P., Johnson, J. W., and Masten, J. M.,** Automated antibiotic susceptibility testing in *Automation in Microbiology and Immunology*, Hedén, G. C. and Illéni, T., Eds., John Wiley & Sons, 1975, chap. 17.
12. **Isenberg, H. D., Reichler, A., and Wiseman, D.,** Prototype of a fully automated device for determination of bacterial antibiotic susceptibility in the clinical laboratory, *Appl. Microbiol.*, 22, 980, 1971.
13. **Chappelle, E. W. and Picciolo, G. L.,**Bacterial adenosine triphosphate as a measure of urinary tract infection, NASA Tech Brief, TSP71-10051, Goddard Space Flight Center, Maryland, March 1971.
14. **Chappelle, E. W. and Levin, G. V.,** Use of the firefly bioluminescent reaction for rapid detection and counting of bacteria, *Biochem. Med.*, 2, 41, 1968.
15. ATP — fast assay for bacteriuria, *Lab. Manage.*, May 1974, 35.
16. **Vellend, H., Tuttle, S. A., Barza, M., Weinstein, L., Picciolo, G. L., and Chappelle, E.,**A Rapid Method for the Determination of Microbial Sensitivity Using the Firefly Luciferase Assay for Adenosine Triphosphate, paper presented at the 74th Annu. Meet. Am. Soc. Microbiol., Chicago, May 1974, 84.
17. *Biometer Flashes*, 760BF5, DuPont Instruments, Wilmington, Del., October 1972.
18. **Delin, S., Monk, P., and Wadso, I.,** Flow micro-calorimetry as an analytical tool in microbiology, *Sci. Tools*, the LKB Instrument Journal, Rockville, Md., 16, 22, 1969.
19. **Binford, J. S., Binford, L. F., and Adler, P.,** A semiautomated microcalorimetric method of antibiotic sensitivity testing, *Am. J. Clin. Pathol.*, 59, 86, 1973.
20. **Russell, W. J., Zettler, J. F., Blanchard, G. C., and Boling, E. A.,** Bacterial identification by microcalorimetry, in *New Approaches to the Identification of Microorganisms*, Hedén, C. G. and Illéni, T., Eds., John Wiley & Sons, New York, 1975, chap. 7.

Chapter 5

PULSE-HEIGHT ANALYSIS (PHA)

I. INTRODUCTION

One approach that can be applied to an instrumented system for microbiological analysis is pulse-height analysis (PHA). This was recognized as early as 1958 by Kubitschek[1] and was the subject of further studies by Curby et al.[2,3] Its application to the detection of bacteriuria was made by Truant[4] in 1962. Further studies by Gall et al.[5,6] have shown that the PHA has the potential to rapidly perform a complete microbiological analysis of a body fluid such as urine including detection, semiquantitation, identification, and susceptibility testing. PHA will be discussed in greater detail than the other approaches to serve as an illustration of the application of the theoretical rules and practical considerations for microbiological instrumentation to a specific system. Pulse-height analysis has been chosen for this detailed discussion because the coauthors of this monograph have had first-hand experience over a period of years in designing, developing, and testing the feasibility of a prototype instrument system applying this approach to the routine microbiological analysis of body fluids.

There is at present no practical automated instrument system commercially available applying the principle of pulse-height analysis to the routine microbiological analysis of body fluids in the clinical laboratory. However, for several years such an instrument has been under development at Sias Laboratories, Lahey Clinic, Brookline, Mass. in a joint venture with Grumman Health Systems Division of Grumman Corporation, Bethpage, L.I., N.Y. U.S. Patents 3,804,720 (1974) and 3,919,050, (1975) have been issued to William A. Curby and assigned to Lahey Clinic Foundation, Inc.

II. PHA

References from the literature relating to pulse-height analysis and experiences and studies with the Curby prototype system will provide the basis for the following discussion.

A. Principle of Action

The microbiological analysis of body fluids using pulse-height analysis is basically an electronic analysis of the number and size of each microorganism in a small sample as it passes through an orifice in an electrical field and produces a pulse which is counted and sorted according to size in a pulse-height analyzer. The properly adjusted PHA can measure pulses through the entire bacterial size range. This analysis is based on the fact that as electrical voltage pulses are produced, one at a time for each microorganism, the bacteria are detected and counted, and the shape and amplitude of the pulse produced is directly related to the volume of the microorganism. The PHA can be used for analysis whenever a sample containing particles is subjected to a test condition which will influence the number, the size, or both of the particles in the specimen.

The PHA applies a physical approach to the microbiological analysis of body fluids which is both rapid and sensitive. The fluid specimen as collected from the body contains various particles including body cells, crystalline chemicals, and living and dead microorganisms. The microbiologist is mainly interested in the living bacteria, presumably growing at the site of an infection, upon which he wishes to perform the analysis.

B. Application of PHA

1. Detection and Enumeration

The detection and enumeration of bacteria in body fluids with the PHA depends on an increase in the number of bacterial-sized particles. To adapt the instrument for this purpose, it is designed to have parameters compatible with the counting and sizing of particles from about 0.3 μm to 4—5 μm in diameter. The detection in a specimen of an increasing number of particles in this size range after incubation indicates the presence of living bacteria, and the rate at which this increase occurs is used as a basis for semiquantitation of the bacteria present in the test specimen.

2. Antibiotic-Susceptibility Testing

Changes in the numbers and/or size of these particles when a test reagent such as an antibiotic is imposed indicates that the test substance has caused a reaction. Antibiotics will inhibit growth when the organism is susceptible, as shown by a lack of increase in total particles when compared to the control, and the size and shape of the bacteria frequently change when the microorganism is susceptible.

3. Identification

The identification of bacteria using antisera can be performed by observing the accumulations of large particles produced by agglutination with homologous antisera, as shown by a shift of the particle distribution pattern to the right (toward the larger particle channels) and an increased count of particles 3μm in diameter or greater coupled with a lack of increase in total particles.

C. Description of an Instrument System and Operation (Curby Prototype)

1. Design Considerations

The application of pulse-height analysis to the microbiological analysis of body fluids in the Curby prototype system is made essentially by using a modified Coulter Counter®* sensor attached to a pulse-height analyzer and a recorder. The Curby system is being designed to be almost totally automated, and the discussion will include not only the description of the features and modifications already incorporated in the prototype instrument system, but also the considerations involved in planning for additional features to be included as this system is automated. It should be stressed again that this discussion is mainly to serve as an illustration of what is involved in producing an automated system for routine microbiological assay of body fluids.

The successful adaptation of the PHA approach to the rapid microbiological analysis of a urine specimen for bacteriuria, and susceptibility and identification of the predominating pathogens in positive specimens, has several requirements, including:

1. Selective differentiation of bacterial size particles in urine
2. Sufficient sensitivity to detect and enumerate 10^5 bacteria per mℓ urine
3. Sufficient accuracy, reliability, and reproducibility to detect a doubling (or less) in the numbers and/or size of bacterial particles as the result of bacterial growth, or in response to added reagents such as antibiotics or antisera
4. Completion of analysis in a short time period, such as 4 to 6 hr
5. Performance of analysis on urine specimens with a mixed bacterial population
6. Production of data that can be clearly interpreted
7. Performance of the analysis at a reasonable cost

* Registered trademark of Coulter Electronics, Inc., Hialeah, Fla.

2. Design Requirements

What is involved in meeting each of these requirements?

1. The selective differentiation of bacterial-sized particles from the larger body-cell debris and the smaller crystalline particles in urine is key to the detection of a bacteriuria, as it is bacteria which increase in urinary-tract infections. This differentiation is principally accomplished by employing a sufficient number of channels to provide adequate separation of peaks produced by particles of significantly different size.
2. The sensitivity to detect and enumerate small numbers of bacteria and the accuracy to detect a doubling (or less) in the numbers and/or size of bacterial particles is dependent in large part on the orifice. Keeping the background "noise" from extraneous particles, electronic interference, orifice clogging, and improper orifice configuration low enough to produce significant data is necessary.
3. Important in a truly rapid performance of a complete analysis is the reaction and reading time for each specimen, which depends in part on the method of analysis chosen and the volume of each specimen read. The capability of the approach chosen to handle mixed cultures, thus eliminating the lengthy isolation procedure, contributes significantly to the overall rapidity of obtaining results.
4. The proper manipulation of the data by the data-processing unit of the pulse-height analyzer brings out the important differences in the control and test results in the clearest form. The selection of the most outstanding differences as criteria for judging the reaction will lead to a definitive interpretation of the results. The presentation of the results by the recording unit also plays an important role in the clarity of the report.
5. Cost-effectiveness in all phases of the design and development of the instrumented system will yield a product cost competitive with existing methods.
6. The capability to fulfill these requirements is seldom attributable to one single factor, but usually is accomplished by the proper combination of a number of factors in the design and development of the instrumented system. For this reason, each component in the system will be discussed separately, and its contribution to fulfilling the stated requirements will be pointed out.

3. Prototype Description

The present prototype system is a semiautomated system composed of three major components including a sensor, pulse-height analyzer, and a recording unit (Figures 1 and 2). This prototype has been shown to be feasible to detect, semiquantitate, identify, and perform susceptibility tests for bacteriuria.[5,6] Each component will be discussed individually including appropriate design considerations.

a. Sensor

In the prototype, the sensor assembly which detects microorganisms in the sample consists essentially of a small orifice placed in an electrical field provided by properly positioned electrodes together with ancillary pumps, a mercury-column metering system, and reservoirs and waste vessels to accommodate the fluid-sample flow. The aliquotted specimens are placed in vials or cartridges for reading.

The details of the sensor assembly to be incorporated in the automated version of this prototype are under study, and the exact design is not complete, as there are many factors being considered.

In designing the sensor, the components are selected carefully to measure most accurately small particles in the 0.3 to 3.0 μm range, which is the size of many bacteria.

FIGURE 1. Sensor unit of Curby prototype of pulse-height analyzer.

The sensor is the heart of the PHA, and the central feature of the sensor is the orifice. Therefore, the consideration of orifice size and configuration is of primary importance. The factors influencing the choice of the orifice design include the orifice dimensions, mounting and clearance.

(1) Orifice

(a) Orifice Dimensions

According to Kubitschek[7] the dimensions of the orifice with respect to diameter and tunnel length, and their relationship to each other, are critical to the accuracy of the analysis. The orifice must be large enough to allow the expected particles to pass through easily, but not so large as to lose sensitivity and definition. The orifice tunnel must be long enough to allow the pulse to mature, but not so long that coincidence occurs in the properly diluted sample. Thus, there is an optimum ratio between orifice diameter and tunnel length dependent upon the particles to be measured. Both aperture diameter and length play a role in the range of particle size that can be determined accurately. For best results, the particle cross-sectional areas should be no more than a percent or two of the cross-sectional area of the aperture for the upper size limit, and the lower limit depends upon electrical noise generated within the aperture or on aperture resistance. A broader range of particles can be counted than can be measured accurately, because sizing is more exacting than counting. The range of particle sizing can be extended by adding an aperture of a different diameter.

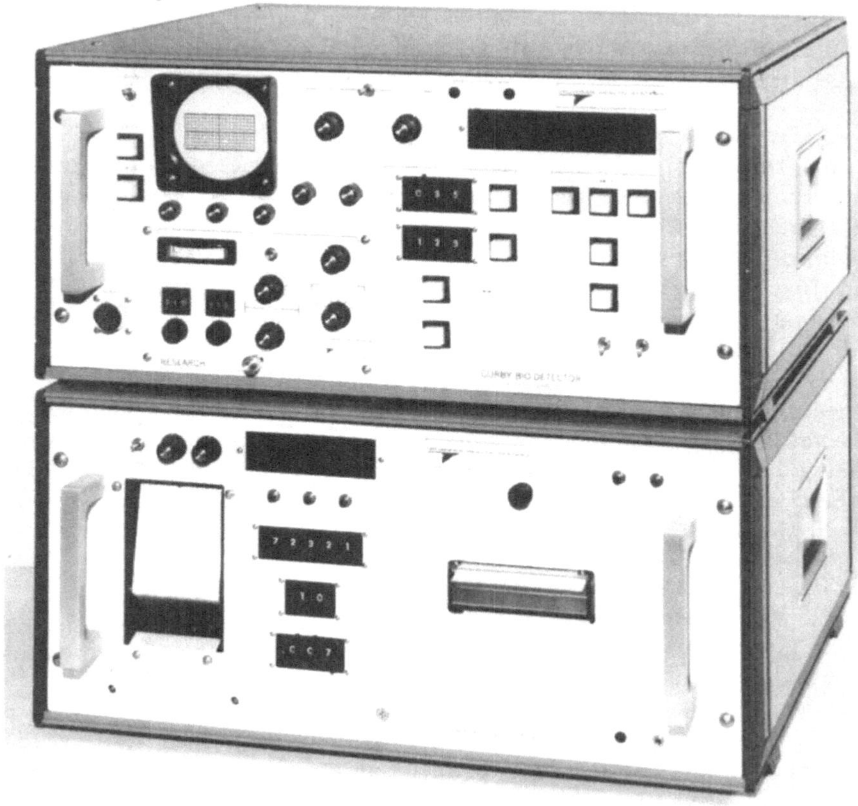

FIGURE 2. Pulse-height analysis and recorder units of Curby prototype of pulse-height analyzer.

To count and size bacteria properly, a small aperture (10 to 30 μm in diameter) is needed, but this small size leads to resistive heating and to blockage of the orifice. The practical size of the orifice is limited by these temperature increases and by the blockages. The choice of orifice size becomes a compromise between the heating and blocking effects and the advantage of a small orifice size on the accuracy of sizing and counting small particles.

The configuration of the orifice with respect to the entrance, tunnel smoothness, and angle of the tunnel should be fashioned carefully to prevent eddying during passage. A poor orifice configuration leads to unnecessary background noise.

(b) Mounting of the Orifice

Other factors affecting the orifice are the material into which the orifice is bored, the mounting, and the device into which the orifice-containing material is placed.

The orifice in its mounting will be placed directly into the sample. Therefore, the materials used, the method of mounting, and the configuration of the mounting must be compatible with the sample composition and size to allow ease of aspiration of the sample. Possible configurations involve mounting the orifice on the side or the bottom of the tube. The size of the orifice tube should bear the proper relationship to the sample size and orifice dimensions to optimize the measurement of the particles passing through the orifice. The orifice mounting selected for use should be geared to be immersed in as small a sample as is compatible with accurate results.

The object of these selections is to have the optimum orifice material and configuration for measuring bacteria-sized particles with the minimum of background noise.

(c) Orifice Cleaning System

The small orifice may become clogged while making a succession of readings. This clogging reduces the orifice size and causes erroneous results, necessitating orifice cleaning. The instrument system should contain an automatic method for recognizing a clog and for removing it by methods such as back flushing. Also to prevent clogging and carry over from the last sample, the orifice should be rinsed and flushed after every reading.

(2) Electrodes

The electrodes used to create the electrical field should be of durable material that will withstand decontamination, be nonreactive, and retain superior properties for maintaining a constant electrical field. Many metals including platinum and nichrome may be used. The configuration, positioning, and sizing of the electrodes should be done properly to ensure their optimum functioning.

(3) Other Components in Sensor

A system of ancillary pumps, a fluid volume metering system (such as a mercury manometer), reservoirs, and waste vessels are needed to aspirate and measure the sample and provide for exact fluid flow rates and disposal.

The mercury metering system as used on the prototype becomes dirty and is then the cause of minor malfunctions of the sensor. The mercury can be cleaned, correcting the problem, but alternate methods for metering are being sought to overcome this difficulty.

(4) Other Considerations

The volume of the sample to be measured is a compromise between the quantity of sample needed for an accurate measurement and the speed of reading the sample. In PHA procedures, 50 $\mu\ell$ is the sample volume often used. It takes about 10 to 12 sec using a 30 μm diameter orifice to read in a well-adjusted system. A smaller sample with a shorter reading time would speed up the analysis. A study to determine the smallest sample compatible with an accurate measurement was performed by Curby et al.[8] It showed that reading times as short as 0.1 sec are practicable. The media carrier or cuvette of this prototype system must also include a cuvette in which the test culture is contained, which does not contaminate other cultures, the operator or environment, and is easily and efficiently disposed of.

It is clear that the sensor and its orifice is key to the production by the PHA of data with the sensitivity, accuracy, reliability, and reproducibility necessary for microbiological analysis, and although details of the sensor for the Curby prototype are still under study, the sensor will probably be patterned after existing instruments with modifications as needed for improvement. Most sensors used in microbiological analyses, such as those described in References 4-9, have orifices 30 μm in diameter in a jewel mounted in the side of a tube. They usually have platinum electrodes and are adjusted by a mercury manometer and necessary pumps to read a 50 $\mu\ell$ specimen. Some also have an automatic flushing system.

b. Pulse-Height Analyzer

The pulse-height analyzer (PHA) contains all the electronics required to count, classify as to size, and store the pulses received from the sensor. The PHA also contains the controls for the operation of both the sensor and recorder. In addition, it has two displays — the Cathode Ray Tube (CRT) which shows the distribution pattern of the particles in the completed test, and the LED numerical display of selected data. Techniques have been developed to analyze the CRT distribution patterns entirely automatically.

The pulses generated by the microorganisms are received by the pulse-height analyzer (PHA) and are sorted and stored by computer into 256 discrete channels according to their amplitude. The PHA determines the run time and displays the pulses on the cathode ray tube, as well as determining the type of record made of each test run.

The PHA unit is involved in several important facets in making measurements and in processing the data, including the accuracy in sizing cells, range of particle-size distribution, and in selection of data for processing.

(1) Accuracy in sizing

Accuracy in sizing particles depends on several factors, according to Kubitschek.[7]

1. Electronic noise levels must be kept low enough to allow the particle signal to be readily measured over and above the noise, providing for good resolution.
2. It is necessary to avoid very large pulses, such as those from contaminating particles or body cells, as these interfere with the measurements of the smaller pulses and prevent proper sizing.
3. Coincidence should be prevented by using a low counting rate so that each particle will give rise to a single pulse proportional to its volume. Too rapid counting yields one pulse from several single particles, erroneously lowering the count and distorting the size distribution in favor of large-sized particles.
4. The balance of aperture lengths, flow rates, and amplifier response must yield pulses of the proper amplitude.

(2) Range of Particle-Size Distribution

The PHA instrument settings can be varied to measure the particles over a wide size range. Since microbiological analysis is aimed at measuring the bacterial-sized particles, the most appropriate range of sizes to be included in the particle distribution pattern are from approximately 0.3 to 3.0 μm. This range may need to be adjusted upward to 4.0 to 5.0 μm to include yeasts and, by providing a means for extending the measuring range electronically, to about 5.5 to 7.0 μm for certain aspects of serological agglutination tests. Factors affecting the range of accurate particle sizing were discussed previously.

(3) Selection of Data for Processing

The PHA generates a variety of data, and a selection must be made of the information pertinent to the microbiological analysis being performed. For example, the data most valuable for detection are those showing an increase of particles, particularly in distribution channels characteristic of bacteria, and a change in the particle distribution pattern. Both show growth of a bacterial culture. An interesting auxiliary measurement is the increase in large particles which seems to parallel growth. The inhibition of growth by antibiotics is determined by a lower total particle count in the test vial compared to the control and, frequently, a change in the particle distribution pattern. Homologous antisera used for identification cause an increase in new, large particles and a general enlargement of all particles. This is seen as a shift of the particle distribution pattern to the right.

For these analyses using the prototype PHA system, the data selected for recording and reduction are (1) the total particle count, (2) the particle count in, and location of, the peak channel, and (3) the particle count in channel 255 (3.0 to 7.0 μm range), as well as a tracing of the particle distribution pattern shown on the oscilloscope.

The PHA module of the prototype instrument system is very flexible and allows a choice of current, gain, and amplitude settings as well as threshold and offset channels to show distribution patterns of particles between approximately 0.3 to 7.0 μm. A balance between the gain and current is determined for each instrument system by calibration of the instrument with latex spheres of known size and remains constant after satisfactory setting during the entire run, usually 300 to 400 ma for gain and 200 ma for current. The threshold setting should be high enough to screen out most of the "noise" created by electronic interference or very small particles, usually about channel 20. The 236 remaining channels will measure particles between approximately 0.3 to 3.0 μm and display their distribution pattern. If particles larger than 3.0 μm are present (such as yeasts or clumped bacteria), the offset may be used to shift the display of the particle distribution pattern to focus on the larger particles. The larger the offset, the larger the particles included in the display. The amplitude setting may be changed as indicated for the system and the particles being measured.

c. *Recording Unit*

The prototype system at the present stage of development has a very simple data recording unit consisting of the Recorder assembly with a paper-tape printer which prints selected data in digital form and strip-chart recorder that traces the oscilloscope display. The data printed on the tape are

- Sample number
- Run number
- Run time (sec)
- Time of test (hr, min, sec)
- Date
- Total count
- Count 255 channel
- Number of peak channel(s)
- Count in peak channel(s)

The particle distribution pattern of the sample just tested is traced on a graphpaper-type strip chart. Both the tape and the chart can be mounted for a permanent record (Figure 3).

Much more extensive data processing systems are possible, including such features as calculation of skewness and standard deviation, indications of particle size (volume) at statistical mean and mode, determinations of volume and population percentiles, and the display of population and volume differential and integral histograms. In addition, they can perform data smoothing and a process that converts particle-pulse amplitudes into a logarithmic basis. Part or all of these capabilities are included in certain models of the pulse-height analyzer already on the market.

The Curby prototype instrument when further developed will carry a more complex data processing system than it does in its present mode, incorporating those features needed to more clearly express the data needed to interpret the results of routine body-fluids microbiological analysis.

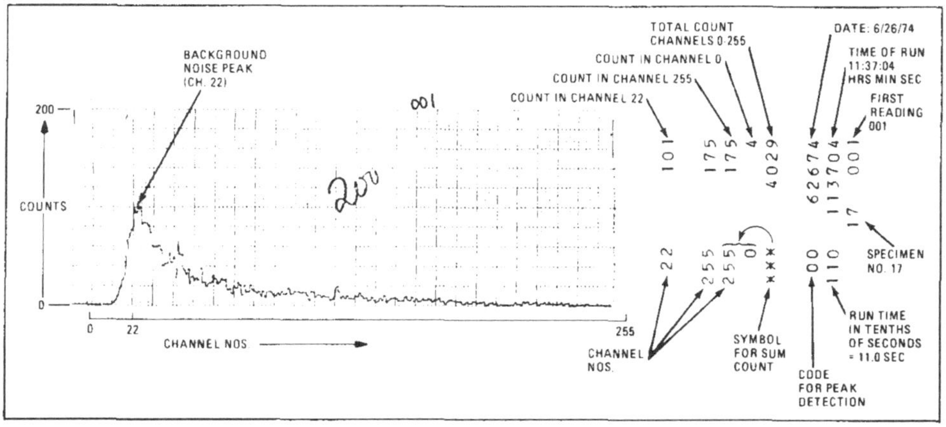

FIGURE 3. Paper-tape print out of test results in digital form with the strip chart recording of the oscilloscope display.

III. OPERATION

The operation of the present Curby prototype is accomplished by the manual dilution of the urine specimen and its placement in a separate incubator-shaker system, manual aliquotting of the specimen and placement of the aliquot on a platform with the orifice immersed in the sample for reading, and removal and disposal of the aliquot after reading. The read cycle is accomplished after the aliquot is positioned by depressing a button to the "ready" position until an adequate vacuum has been established. This is followed by depressing the "volume run" button, which initiates and stops the reading of the volume of the aliquot measured by two photoelectric cells. The data are recorded automatically, and on command, the instrument will record the manually located peak channel of the distribution pattern and graph the oscilloscope display. Data manipulation and interpretation are manual.

To perform an antibiotic susceptibility test or a serological identification when a positive bacteriuria has been detected, the recently detected culture is manually diluted, and antibiotic or antisera are manually added to separate aliquots of the cultures. The tests are manually placed into the incubator-shaker and read as described above at the appropriate time.

From this description, it is evident that the present prototype of the Curby PHA is a semiautomated instrument system consisting of a manually operated sensor and an automated pulse-height analyzer and recorder and display. The data must be catalogued and interpreted manually. It is expected that the next generation prototype will be more completely automated, taking into account the various theoretical and practical aspects discussed in this monograph. This automation will center around the front end or sensor to enable the specimen to be incubated and read without manual intervention. Because of the many readings on each specimen necessary to develop the dynamic data needed to interpret a reaction, an instrument system using pulse-height analysis is not suitable for routine microbiological analyses in a clinical laboratory unless the reading sequence is automatic. This has been amply demonstrated by the fact that although the feasibility of using pulse-height analysis for bacteriuria determinations has been recognized for 15 years,[4] and later confirmed in other laboratories by Gall et al.[5, 6] and Smither,[9] the unautomated instruments (Coulter Counters®) on which this capability was demonstrated have not been used routinely for this purpose.

The integration of the data related to the selected criteria and the interpretation of the data also has a high priority in the automation plans.

Whether or not automation should be used to handle the decisions about which antibiotics or antisera to use or for the transfer and dilution of positive bacteriuria specimens being further tested remains to be decided. The decision will be based on trade-offs involving difficulty of automation, cost, desirability, etc. The automation of the sample preparation is probably not warranted as the specimens arrive at the laboratory in a variety of containers, making an automated interface difficult, while the manual dilution of the specimen is a simple procedure. The degree of automation of the pulse-height-analyzer system thus becomes a trade-off between feasibility and desirability.

In designing the PHA instrument system, the fact that the analysis is being performed on living, potentially hazardous particles imposes certain safety restrictions, especially with respect to the culture containers. All of the considerations concerning culture containers discussed in the section on "Practical Aspects", which includes lack of toxicity, configuration, volume, discharge or uptake of specimen without contamination of the sample, environment or operator, and type of material, apply to the PHA system. In addition, the freedom of the counting media and container from small particles is essential.

IV. METHODS FOR MICROBIOLOGICAL ANALYSES OF URINE SPECIMENS

Feasibility has been demonstrated for use of the PHA prototype instrument system in performing a complete microbiological analysis, including detection, semiquantitation, identification, and susceptibility testing (Gall et al.[5,6]).

Certain considerations in the development of the methods for use with the PHA are common to most or all of the instrumented approaches and have been discussed in Chapters 2 and 3. These relate to theory and to test culture, test method, and data collection and handling. In addition, there are several special factors involved in the procedure employing the PHA for urine specimens.

A. Considerations

1. Concentration of Cells in Inoculum

The PHA gives the most accurate results with the least coincidence when the particle count per 50 $\mu\ell$ sample is less than 30,000. For this reason, specimens are prepared with an estimated particle count low enough that normal cultural growth will not bring the particle count of the control above 30,000/50 $\mu\ell$ during the test period. In practice, initial counts between 2000 and 5000/50 $\mu\ell$ usually are desirable. Most urine specimens fall within this range when diluted 1:250 (although heavily infected urines may give a higher count). Usually, the detection of a bacteriuria can be accomplished at or below 30,000 counts/50 $\mu\ell$.

2. Volume of Sample

There are several aspects to the consideration of the volume of the sample. All are important in establishing the method and in instrument design. Included are the total volume of the sample placed in the instrument, the volume of the aliquot taken for reading, and the volume of the sample aspirated through the orifice at each reading.

The volume of the aspirated sample is important in establishing the reading time of each sample. Therefore, the number of samples that can be read per unit time is based on the volume of the aspirated sample plus any ancillary procedures performed at each reading. To allow the most efficient utilization of the instrument, the smallest volume compatible with accuracy should be read at each time period.

The volume of the aspirated sample at each reading will also influence the volume of the aliquot of the total sample taken at each reading. The volume of each aliquot multiplied by the number of readings will influence the total volume of the specimen placed in the instrument, which in turn, affects the total capacity of the instrument.

Thus, the volume of the aspirated sample eventually influences the capacity of the instrument and the length of time required to process a unit sample. This, plus the time interval between the start and completion of the test analysis, determines the number of specimens that can be processed in a day per unit instrument system.

3. Time Interval Between Readings

Readings should be made at zero time with the PHA procedure to establish the particle background, including the total and 255 channel count and the particle distribution pattern. The frequency of additional readings will be dictated by the incubation time anticipated to produce a change in the zero-hour results. For optimum results, there should usually be at least three readings on each specimen at different reading times corresponding to anticipated changes in the background readings of positive specimens. Certain parts of the analysis, such as detection (because the culture must progress from lag to log growth), will take a longer time to develop changes in the background measurements than the susceptibility or identification tests that are started immediately after detection with cultures already in the log growth phase.

The time of the final reading may also be important. Some body fluids are normally sterile, such as blood, spinal fluid, and aseptically aspirated urine, and the presence of any bacteria constitute an infection. However, other specimens, such as clean-catch urine, usually are contaminated to some extent with normal body flora during collection. The PHA is nondiscriminating and detects contaminants as well as the infecting bacteria. It is, therefore, necessary to establish a procedure to separate the contaminants usually present in lower amounts from the more predominant pathogens. This is done by timing the readings so that the more numerous pathogens will reach the level of sensitivity of the instrument required for detection, but the contaminants present in smaller numbers will not have reached the level of detection. Thus, the detection of the bacteriuria is time-dependent and the timing of the last reading must correspond to the time when the predominant pathogens may be expected to reach the detection level, but the contaminants have not.

4. Temperature

As discussed in Chapter 3, almost all tests involving human clinical specimens are incubated today at 37°C (body temperature), and this incubation temperature is used with the PHA. In addition, it is necessary to maintain the aliquot of the culture at 37°C during reading of the bacteriuria tests as crystals from the urine dissolved at 37°C may reappear upon cooling and increase the total count, thus erroneously mimicking one aspect of growth, i.e., an increase in total particle count.

5. Reagent Broth

The broth in which the culture is grown must be prepared with several requirements in mind, including broth composition and particulate background.

The broth for use in the PHA must have a low particle count and also be an electrolyte so that it will conduct current between the electrodes. It must also support adequate growth of the bacteria expected to be found in the body fluids under test. In general, a rich broth is best, but there is also a limitation on the carbohydrate content, as too much of certain carbohydrates may cause the production of slime by some bacteria which is not desirable as the small orifice may become plugged by this substance. The broth must also be compatible with other test reagents, such as antibiotics, which may be used together with this broth.

B. Application to Microbiological Analysis of Urine Specimens

The method described here is the application of the PHA to microbiological analysis of urine for bacteriuria using the prototype PHA. This determination was chosen because bacteriuria is one of the most common microbiological examinations performed on body fluids. Bacteriuria is usually defined as urine having 10^5 or more significant bacteria per mℓ of clean-catch urine (Kass)[10], but it is recognized that clean-catch urine may be contaminated with normal body microflora in less than 10^5/mℓ, usually 10^3/mℓ or below.

The initial problem of detection of a bacteriuria then becomes not only the detection of bacteria, but of distinguishing the predominant pathogen from the contaminating body bacteria, i.e., semiquantitation. This can be done by reading the sample until bacteria present in the specimen at 10^5 have been detected but before the contaminant, present in lower numbers, has reached the detection level. In general, the microorganisms commonly causing a bacteriuria reach detection level in 2½ hr, while contaminants require 4 hr or more.

1. General Method of Detection and Semiquantitation

The first task of the microbiologist using the PHA in bacteriuria tests is to detect living bacteria in the specimen, i.e., to separate the viable bacterial particles from the other extraneous particles. The key to this separation is the unique ability of viable bacteria to reproduce in a suitable fluid medium held under appropriate conditions. The method for detection of the bacteria is thus straightforward. The instrument is set up to detect particles of bacterial size and the specimen is inoculated into a suitable medium after having been freed of as many extraneous particles and threshold inhibitors as possible, usually by dilution. The zero-hour reading is made on this suspension to get a background particle count and a particle-size-distribution pattern by passing a small quantity of the suspension through the small orifice in an electric field. A period of incubation is followed by further readings during which viable bacteria may multiply, while the extraneous particles will remain unchanged. Bacterial presence is detected when the total number of particles increases and a pile-up of approximately the same-sized particles appears on the distribution-pattern display on the oscilloscope. This is a well-defined peak in a relatively narrow band of channels. The peak channel indicates the average size of the bacteria, a characteristic which is often typical of a certain bacterial type. If there is only one predominating bacterium, one peak will occur. If there are two or more predominating bacteria, there will usually be a corresponding number of peaks. The exception to this are streptococci, which in pure culture form multiple peaks, characteristically harmonics of each other. These probably represent pairs, fours, and short chains.

Criteria for detection of a positive bacteriuria specimen are that within 2½ hr there occur:

1. A substantial increase in total count (usually by 2 times)
2. An increase in channel 255 count by 2 times
3. An emerging particle distribution pattern

The doubling of the total count indicates at least one generation of growth. When the background count at the zero-time reading is very high, due to debris or media with too many particles, a true doubling of the count may not occur at the reading when the other two criteria appear, but there should be at least a substantial increase in total count. If in doubt, repeat the reading in 15 min. The total count will always increase in subsequent readings if living bacteria are present.

The increase in the large particle count in channel 255 is an interesting phenomenon. It may be due to the increased size of cells immediately before division or the tendency of the divided cells not to separate immediately after division. The formation of slime may also account for it. The exact cause of the doubling of the count in channel 255 awaits further study, but it is a constant feature of bacterial growth in the reagent broth used in these tests and is a very reliable indicator of bacterial growth.

The emerging particle distribution pattern is the most reliable indicator of bacterial growth. The specimen of urine at the zero-time reading always shows some pattern indicating the presence of particles, usually peaking below channel 30. This is referred to as "noise". It is probably composed of very small particles and electronic interference. Negative specimens develop no different pattern on further incubation, but those containing live bacteria develop a peak in the channel corresponding to their average size, usually between channels 45 to 150. This emerging pattern is easily seen at the usual instrument setting with all bacteria, but yeasts may be too large to appear in any of the 254 channels normally used. However, the other two indicators will alert the operator to the presence of a microorganism as the total count and the count in channel 255 will increase. By switching the offset 100 channels, a typical distribution pattern will then be brought into the picture for the large-celled yeasts.

The time limit of 2½ hr for the appearance of the three criteria for a positive bacteriuria was established by correlation with traditional methods. 97% of the urine specimens showing a colony count of 10^5/mℓ or greater also developed the three criteria within 2½ hr. Apparently within 2½ hr, 10^5 bacteria per mℓ of urine were able to multiply enough to produce a particle distribution pattern in the bacterial "window", double the count in channel 255, and double the original background count. This is somewhat amazing as, following the procedure to be described and allowing for dilutions and small sample read, only 20 microorganisms will be measured in the zero-hour reading of 50 mℓ of specimen, containing just 10^5 bacteria per mℓ.

When these three criteria are fulfilled within 2½ hr, a bacteriuria has been detected. Although one or two criteria may give a good indication of the presence of living bacteria, three criteria are required for optimum results. This gives us three of the criteria needed to fulfill the engineering and biological rules used to design the instrument system.

2. *Specific Method for Detection and Semiquantitation*

The specific method established to detect bacteriuria in urine specimens using the prototype PHA is based on the theoretical and practical considerations discussed previously.

Immediately upon receipt of the specimens in the laboratory, the urines are inspected and cloudy specimens are shaken by hand for 10 sec in a 50°C water bath to effect clearing if the cloudiness is due to soluble crystals.

Each specimen is then diluted aseptically 1:250 times in particle-low reagent broth which has been warmed to 37°C. The diluted urine is mixed and an aliquot is poured into a vial preheated to 37°C and read immediately on the prestandardized PHA. The remaining diluted urine is placed in a 37°C incubator-shaker, with the shaker set at 150 rpm, and incubated for 1½ hr.

The first reading (001) is made on a 50 μℓ sample immediately after the urine has been diluted and is intended as a zero hour or background count.

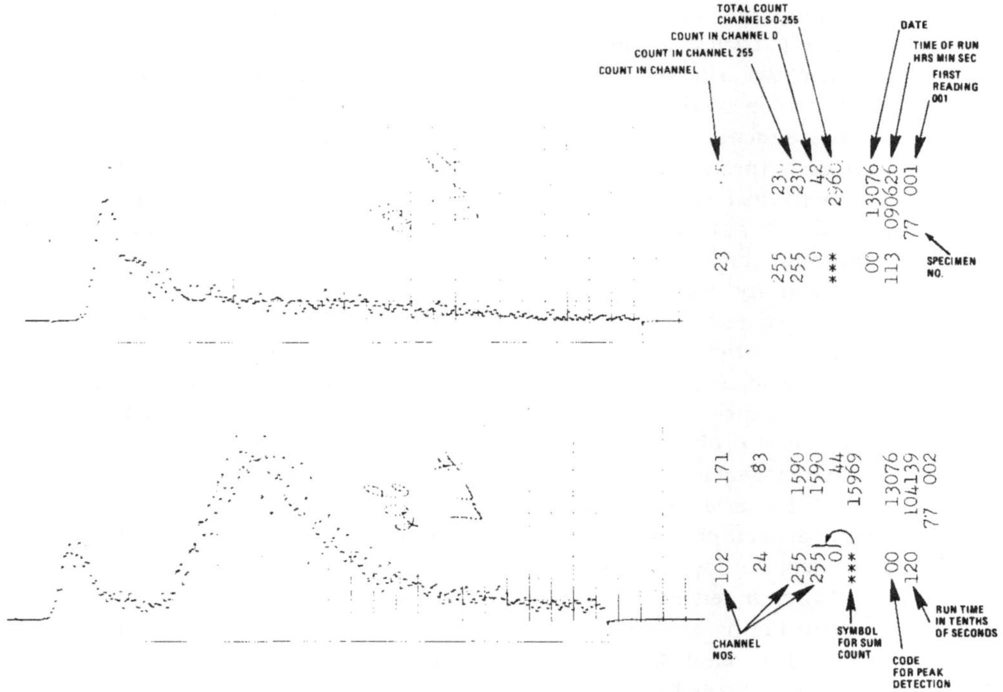

FIGURE 4. Test record of a urine specimen positive for bacteriuria.

After an 1½ hr incubation, each sample is read again (002). By this time most of the specimens positive for bacteriuria have emerged. If necessary, a third (003) reading is made at 2½ hr. The specimens are judged positive using the following criteria, one or all of which may be manifested by the data generated during the reading of the 50 $\mu\ell$ sample of specimen (Figure 4).

1. The emergence of a pattern different from that shown in the zero-hours reading, i.e., a pattern other than a "noise" pattern in channels 20 to 29 is one criterion. Occasionally, a true pattern emerges in the zero-hour reading when the specimen is strongly positive. This pattern is always amplified in the 1½ hr reading, if the bacteria are viable.
2. An increase in the channel 255 count by at least 2 times.
3. An increase in total count usually by 2 times.
4. These criteria must be evident at least by the 003 (2½-hr reading).

Those specimens that do not fulfill at least two of these criteria after a 2½-hr incubation are judged to be negative; thus the exact timing of the readings is very important. Those specimens which are judged to be positive are prepared for antibiotic-susceptibility testing and identification by antisera. These tests are done in parallel.

3. General Method for Antibiotic Susceptibilities

The antibiotic susceptibility test is performed by aliquotting the just-detected standardized culture in log phase and adding the antibiotics to each vial individually while maintaining a control vial. A period of incubation is followed by the reading of the control and each test vial (Curby).[11] The control and cultures resistant to the added antibiotic will show an increase in total particles and in the count in the peak of distribution pattern. Susceptible bacteria, as compared to the control, will show fewer total particles and a lower count in the peak-of-distribution pattern.[5] Each peak in a mixed culture will react independently, while in contrast, all of the multiple peaks in the pure *Streptococcus* culture will react in the same way to the test antibiotic, i.e., all peaks show resistance, or all show sensitivity. This acts as a cross-check in establishing a pure vs. a mixed culture if multiple peaks occur in the detection pattern.

The antibiotic-containing tests are compared to the control, and susceptibility to the antibiotic is indicated when:

1. The total count is substantially lower than the control
2. The count in the peak channel is lower
3. Particle distribution pattern changes (usually)
4. Change in 255 channel count (usually)

If these changes do not occur, the culture is considered resistant. A resistant culture is usually nearly identical with the control.

The lowered total count and lower count in the peak channel reflects the slower, or lack of, growth of the culture in the presence of an antibiotic to which it is susceptible. Certain antibiotics, while inhibiting bacterial division, cause the organism to increase in size. This shifts the particle distribution pattern to the right and also usually increases the count in channel 255. Other antibiotics may reduce the size of the bacterial cell and shift the pattern to the left. This usually decreases the 255 channel count. Some antibiotics do not change the size of the susceptible bacterium, resulting in a peak in the same channel as the control.

4. Specific Method for Antibiotic Susceptibility

The specific method to perform susceptibility tests based on a consideration of theoretical and practical aspects of instrumented microbiological analysis is described below. The positive detection culture is diluted with reagent broth to 5,000 counts per 50 μl ± 1000 counts and placed in containers. In the event that the positive specimen does not require dilution, the undiluted specimen is placed directly into the sterile containers.

Judged by a particle distribution pattern which is different for Gram-negative rods and Gram-positive cocci, appropriate antibiotics are chosen and antibiotic impregnated discs are placed into each container with one reserved as a control. A check of the correctness of the dilution is made by taking the total count of the control (020). If necessary, adjustments in the dilution are made prior to adding the antibiotics and the 1 hr incubation at 37°C on an incubator-shaker.

Following incubation, a reading is taken of the control culture (021) and if the total count has at least doubled, a reading is taken immediately of the antibiotic-containing cultures. The susceptibility of the culture to each antibiotic is judged by comparison of the data from the 021 (control) reading and the antibiotic-containing cultures. The culture is judged to be susceptible when there is (Figure 5)

1. A lower total count
2. A lowering of the count in the peak channel
3. An alteration in the particle distribution pattern (usually)
4. Frequently, a change in count in channel 255, depending on the test antibiotic.

If the control has not doubled in 1 hr (021 reading), the control and antibiotic cul-

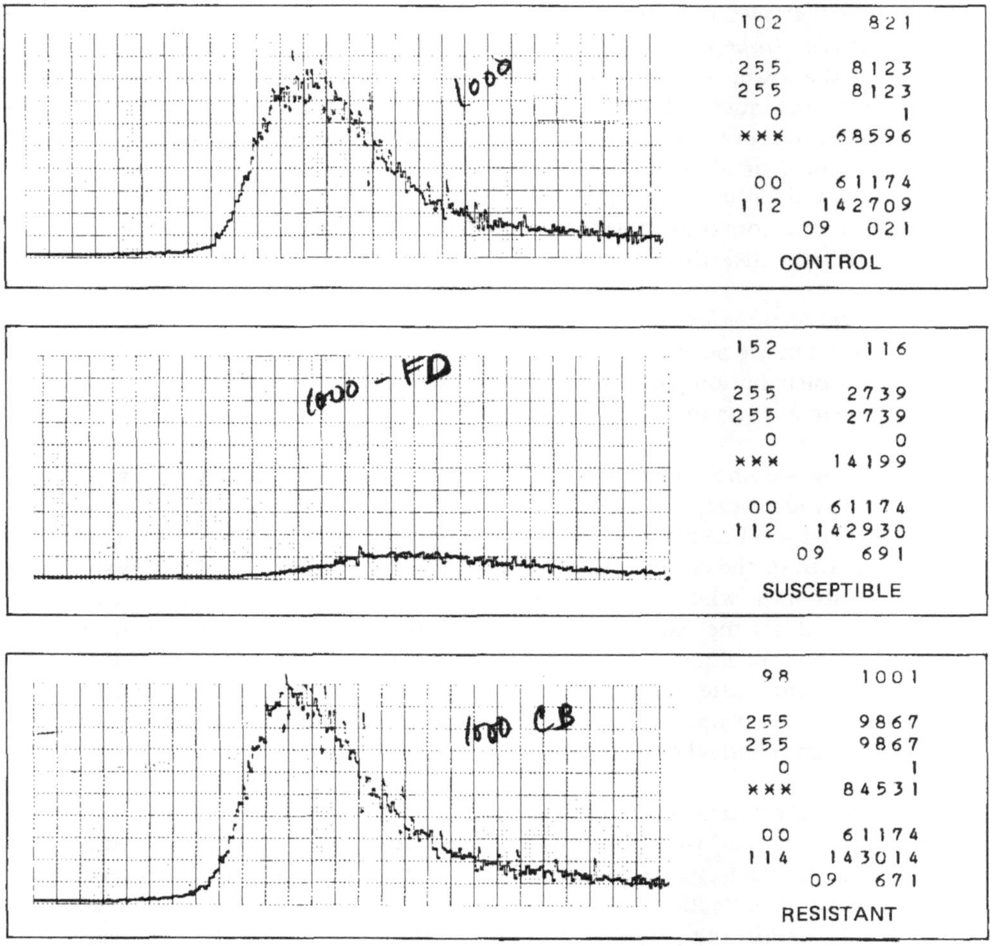

FIGURE 5. Test record of a bacterial culture of *Proteus mirabilis* showing susceptibility to furadantin and resistance to carbenicillin.

tures are reincubated for ½ hr and read again (022). It rarely occurs that the 1 hr reading shows too little growth.

The typical sensitive reaction shows a change in all criteria with the possible exception of the peak shift and 255 channel count which are antibiotic dependent. A resistant bacterium shows little or no difference in data from that in the control.

5. General Method for Identification by Antiserum Agglutination

Identification of bacteria by agglutination using homologous antiserum is a very specific procedure which is well suited for use with PHA (Gall et al.).[6] The reaction between the bacteria and the homologous antiserum results in the formation of particles larger than the bacterium and eventually, clumping occurs. Both reactions are readily detected with PHA. The procedure using the PHA, therefore, is not new, but substitutes a very sensitive instrument for the human eye, which can detect only gross clumping. The method is rapid and simple and can be performed in parallel with the antibiotic-susceptibility test.

A portion of the diluted culture prepared for susceptibility testing is used for identification by adding various antisera, shaking briefly, incubating at 37°C, and then reading after a brief period of standing at 37°C. The heterologous antisera and the control will produce a similar increase in particles and a distribution pattern peaking in the same channels, while the homologous antisera will cause a lower particle count, a shift of the peak to the right, and an accumulation of large particles. As with the antibiotic-susceptibility test, each peak in a mixed culture will act independently while all the peaks of a pure multipeaked culture will react in the same way to the homologous antiserum. The reaction of different peaks to different antisera confirms the presence of a mixed culture. To avoid having an unwieldy number of antisera to test, pools of antisera can be made and the individuals in the reacting pool can then be tested individually. A prediction which aids in antiserum selection can sometimes be made for the type of organism present based on rapidity of growth and the characteristic peak channel.

The antiserum-containing tests are compared with the control and a positive identification is made when:

1. The count in the peak channel of particle distribution pattern is lowered.
2. The count in channel 255 is increased proportionately.
3. The peak channel shifts to the right (to a larger number).
4. The total count has not increased as much as the control.

The changes in the particle distribution pattern are the most reliable index of reactivity. There are varying degrees of reactivity, and the strongest reaction flattens the peak channel to near the baseline, shifts the peak channel above 255, and allows no increase in growth. The lowering of the count in the channel and the total count indicate the smaller number of particles present compared to the control. The shift of the peak channel to the right and the greater accumulation of particles in the 255 channel indicate the formation of clumps of agglutination of the bacteria.

To aid in the interpretation of the test results, the data from the 255 channel are refined by a calculation to show the proportion of large particles (channel 255 count) which accumulate during an antigen-antibody reaction based on the total particle count. The calculation is necessary because the large particles which accumulate during the test period may be related to the growth of the culture or to the antigen-antibody reaction, or to both if an antiserum has been added. In reality, the actively growing control, to which no antiserum has been added, accumulates large particles in channel 255 directly in relation to the growth of the culture, as expressed by the total count. The antigen-antibody reaction also accumulates large particles. In an attempt to show how much of the accumulation is due to the antigen-antibody reaction and how much to growth, the accumulation (count in 255 channel) is divided by the total count to calculate the proportional accumulation. Then the proportional accumulation of the control (which is due to growth) is subtracted from the proportional accumulation of each test (which is due to growth and possible antigen-antibody reaction). This gives the "excess" accumulation, or that due to the agglutination reaction only. All ratio values are expressed × 100.

A simple confirmation of the agglutination reaction may be performed using the offset capability of the instrument. By offsetting the particle distribution pattern by 250 channels, the size of the large particles accumulated in the agglutination reaction can be detected and measured since they usually form a peak at 200 in the offset pattern (actually channel 450).

6. Specific Method for Identification by Antisera

The method based on the theoretical and practical aspects discussed is described below.

The steps in the procedure follow each other in rapid sequence. As soon as the culture is judged to be positive for bacteriuria, a portion of the culture which was diluted to 5,000 counts per 50 μl for the antibiotic-susceptibility tests is used for the serological-identification tests. These two tests are run in parallel.

The culture is incubated a short time, such as 5 to 10 min, and read again on the PHA to insure that the culture has not dropped out of log phase of growth. If the culture is in lag, this is judged by a lowered total count, smaller peak-channel number and count, as well as a lower 255-channel count. If the culture has dropped into lag phase, it is reincubated for several minutes and then read as before.

If the culture is in the proper phase of log growth, it is divided into several vials. One is retained as a control, while the other vials are used for a "screen" test. Each one contains an appropriate concentration of different known antiserum or groups of antisera reactive against a common bacterium or pathogen. The antisera to be used are chosen based on the peak-channel number and growth characteristics which are often typical of certain bacterial species. The exact concentrations of antigen and antibody used in this test are variable, but one advantage of this procedure is the capability to add an antibody excess and to allow the antigen to grow to the proper proportions for the agglutination reaction. With the usual inoculum of test culture, 1:200 to 1:1000 dilutions of homologous antiserum in the test vial usually give a strong reaction within 20 min.

The control- and screen-test vials are swirled and shaken gently for a short time and are then allowed to stand several minutes before reading on PHA. Usually, several more readings are made at appropriate intervals of 10, 20, 30, and 45 min from the time of inoculation.

The data are observed for a change in the particle distribution patterns such as a shift in peak and peak-channel number and count, as well as the 255 channel and the total count. The proportional accumulation of large particles in each vial is calculated and the screen-test results are compared with the control. If desired, the instrument may be offset 250 channels and the offset peak determined in the new "picture" for both the control and the tests. This is done so that particles originally falling in channel 255 or above are now shifted to occupy channels 1 to 254. Therefore, channel 200 would, after the shift, contain particles large enough to occupy channel 450 if the original display had not been cut off at channel 255.

Using the values from the control vial as a basis for comparison, the data from the test-antiserum-containing vials are examined for evidence of clumping, which include (Figure 6):

1. Shift in peak channel number to the right
2. Lowering of the count in the peak channel
3. Absence or lack of significant increase in total count
4. Greater proportional accumulation of large particles in channel 255
5. Greater accumulation of large particles in the 200 channel of a 250-offset reading (450 channel)

A test vial showing these criteria is considered to contain antiserum homologous to the test culture, and an identification is made. If the antiserum in the vial was a pool of antisera, the test is repeated with each individual antiserum represented in the pool.

These data are in many ways interdependent, expressing the same reaction in differ-

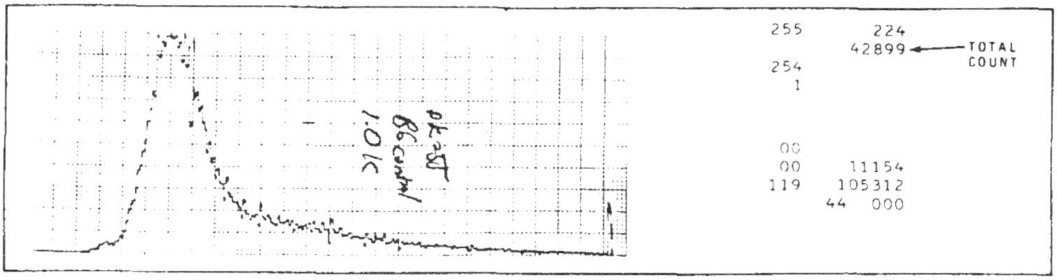

E. Coli, 026-B6 - Control - No Antiserum Added

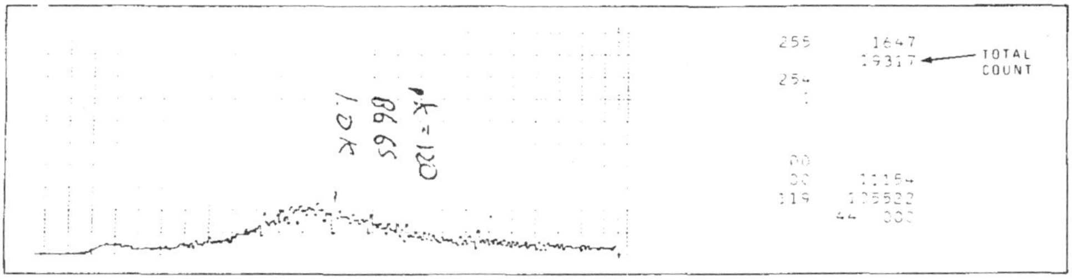

E. Coli, 026-B6 - Homologous - Antiserum 026-B6 Added

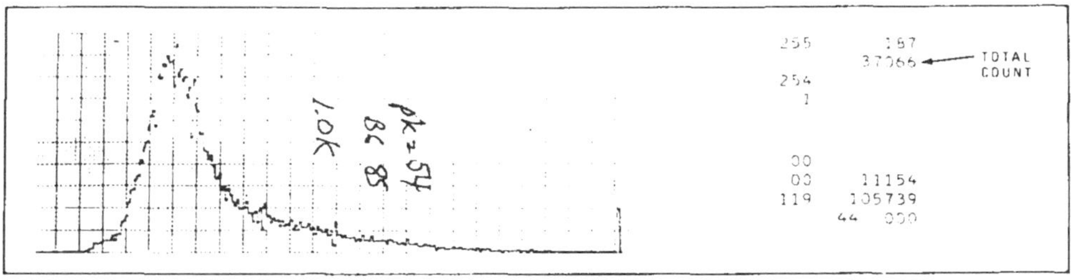

E. Coli, 026-B6 - Heterologous-Antiserum 127-B8 Added

FIGURE 6. Homologous antiserum reaction between *Escherichia coli* 026.B_6 and antiserum for *E. coli* 026.B_6 and heterologous antiserum for *E. coli* 127.B_8 showing no reaction with the *E. coli* 026.B_6 culture.

ent ways, and serve as a cross-check in interpreting the data, which is important in all analysis, but is doubly important in detecting mixed cultures.

V. INSTRUMENT FEATURES

The application of pulse-height analysis to microbiological analysis of body fluids has several advantages.

1. The time interval between collecting the specimen and reporting the results of the analysis to the doctor is of prime importance in the effective treatment of the patient. In the traditional methods, the prolonged period of incubation (12 to 24 hr) necessary to purify the cultures is the chief impediment in obtaining rapid results. The advantage of PHA is that the purification procedure can be eliminated because the instrument does not require a pure culture, as most bacteria form characteristic peaks. Thus, two predominating bacteria, both present in substantial number, form two separate peaks in about 2½ hr. Susceptibility and serological

tests are based on applying the criteria to each peak separately and require one additional hour of testing.

2. A complete bacteriuria analysis can be completed in about 4 hr, including detection and semiquantitation, identification of the predominating pathogens, and susceptibility testing.
3. The criteria to indicate the analytical data are clear-cut, and the fact that several criteria are used for the interpretation of the data allows a cross-check on the individual data points to establish its reasonableness.
4. PHA is a very sensitive measure of reactions occurring in the test vessels.
5. Both aerobic and anaerobic cultures can be tested.

VI. INSTRUMENT REQUIREMENTS

1. The test requires particle-low broth and specially rinsed particle-low test vials. The specimen should also be particle-low in the bacterial size range except for the infecting bacteria.
2. The testing environment should be reasonably free of excessive dust and electrical interference; a normal room temperature and humidity should be maintained.
3. As discussed previously, the ability to attain and maintain log growth of the test culture is needed to obtain optimum results.
4. For fully automatic handling, a cuvette has been designed and will be required for this instrument system.

VII. MARKET STATUS

The instrument, which has not been tradenamed, is in the prototype stage, and introduction on the market is indeterminate. It has not entered testing for FDA approval.

VIII. PRACTICAL CONSIDERATIONS

There are several practical considerations related to the design and use of a pulse-height analyzer.

A. Counting Solutions

The solution in which the bacterial particles are suspended for analysis on the PHA must be an electrolyte which is low in bacterial-sized particles, has a low electronic noise, proper osmotic pressure, and good growth-promoting qualities. The chosen growth medium should offer a compatible combination of these qualities.

B. Cell Counting

Different problems are encountered in accurate cell counting when the counts are low or high.[7] At low counts, with cell volumes of 1 to 2 μm^3, particulate background counts contribute most of the error, but as the cells fall below 1 μm^3, electronic noise contributes heavily to the background count. When the concentration of cells increases, background particulate matter is of lesser consequence, but coincidence plays a larger role in the accuracy of counts. Coincidence occurs when two particles are in the orifice at the same time but are recorded as one count. Thus, coincidence causes underestimation of the true count. It is negligible below 5000 counts/sec, but increases rapidly with an increase in cells.

C. Sample Container

The sample container and its configuration are very important in planning for automation. The sample may need to be manipulated in its container several times, especially when a complete analysis is performed with the PHA. When the sample is read several times in contact with a common probe, it may be necessary to aliquot the sample to prevent sample contamination. This means the container will discharge some fluid and then have to reseal itself. Intermediate dilutions may also be carried out in the container. Therefore, some measurement of volume will be necessary, and the container will have to be large enough to accommodate the required amount of diluent. The overall configuration of the vessel must be compatible with automation. The material from which the container is made will require sterilization and must be reasonably particle free. Therefore, selection of the plastic for the container is critical.

IX. GENERAL COMMENTS

Curby's prototype instrument system employs the pulse-height analyzer to perform complete microbiological analysis of urine for detection of bacteriuria, antibiotic susceptiblity, and identification of the predominating pathogens in about 4 hr. The results of a limited trial on 679 clinical specimens of urine have been compared with classical methods of analysis and show the following agreement:[5]

Detection, 97%
Antibiotic susceptibility, 92%

Preliminary tests on 79 bacteriuria specimens for serological identification gave 90% accurate results.[6] These tests were performed with a limited spectrum of antisera, and it is anticipated that these results will be improved when a broader spectrum of antisera is used.

Other studies by Truant et al.[4, 12] using the Coulter Counter® to detect a bacteriuria have shown promising results.

X. OTHER PHA INSTRUMENT SYSTEMS

There are two fully developed pulse height analyzers on the market for performing research-type microbiology, but which are at present impractical for routine microbiological analysis of body fluids because the nonautomated sensors require time-consuming, manual, reading operations.

There are two firms offering these pulse heights analyzers.

Coulter Electronics, Inc. of 590 W. 20th St., Hialeah, Fla. 33010 has a series of models of Coulter Counters® such as Models ZB, ZBI, and ZH to which a Coulter Channelyzer® Model H4 is attached to analyze the size distribution of particles counted. Coulter Electronics, Inc. holds many patents covering this system, and offers a bibliography listing over 175 papers reporting the use of the Coulter Counter® for microbiological studies; however, less than five reported use in a routine clinical diagnostic procedure, mainly for bacteriuria.

Particle Data, Inc., P.O. Box 265 (111 Hahn St.) Elmhurst, Ill. 60126 has tradenamed their pulse-height-analysis system Electro Zone/Celloscope®. It is intended for particle count and size analysis and has a patented feature related to the logarithmic output (U.S. Patent 3,626,166).

Pulse-height analyzers produce the same basic data, but the instrumented systems on the market differ from the Curby prototype principally in the number of statistical techniques performed on the data, thus facilitating the comparisons between distribution curves. These systems, depending on the system and model, may offer differential or integral histograms, calculations of skewness and standard deviation, indications of particle size (volume) at statistical mean and mode and determinations of volume and population percentiles. Additional manipulation of the data edits and extrapolates data curves, smooths the curves, converts to area- and volume-distribution data, integrates cumulative curve forms, and takes differential profiles between reference and unknown samples. Particle-pulse amplitudes may be converted to a logarithmic basis, as with Particle Data's Electro Zone/Celloscope®. This treatment is generally considered to be best for data interpretation, since normal size distributions of particles are usually log-normal.

The chief purpose of this chapter is to illustrate the many considerations involved in producing an automated instrument system.

REFERENCES

1. **Kubitschek, H. E.**. Electronic counting and sizing of bacteria, *Nature (London)*, 182, 234, 1958.
2. **Swanton, E. M., Curby, W. A., and Lind, H. E.**, Experiences with the Coulter Counter in bacteriology, *Appl. Microbiol.*, 10, 480, 1962.
3. **Curby, W. A., Swanton, E. M., and Lind, H. E.**, Electrical counting characteristics of several equivolume micro-organisms, *J. Gen. Microbiol.*, 32, 33, 1963.
4. **Truant, J. P., Brett, W. A., and Merckel, K. E.**, Application of an electronic counter in the evaluation of significant bacteriuria, *Henry Ford Hosp. Med. Bull.*, 10, 359, 1962.
5. **Gall, L. S. and Curby, W. A.**,Instrumented approach to microbiological analysis of urine specimens, presented at the 76th Annu. Meet. Am. Soc. Microbiol., Atlantic City, N.J., 1976, 40.
6. **Gall, L. S., Baum, J. M., and Curby, W. A.**, Serological technique for bacterial identification using a particle counter, presented at the 78th Annu. Meet. Am. Soc. Microbiol., Las Vegas, Nev., May 1978.
7. **Kubitschek, H. E.**, Counting and sizing micro-organisms with the Coulter Counter®, in *Methods in Microbiology*, Vol. 1, Norris, J. R. and Ribbons, D. W., Eds., Academic Press, London, 1969, chap. 17.
8. **Curby, W. A., Swanton, E., Wallace, J., and Lind, H. E.**, Determination of the minimal sampling interval for electronic counting of microorganisms, presented at the 64th Annu. Meet. Am. Soc. Microbiol., Washington, D.C., 1964, 24.
9. **Smither, R.**, Rapid screening for significant bacteriuria using a Coulter Counter, *J. Clin. Pathol.*, 30, 1158, 1977.
10. **Kass, E. H.**, Bacteriuria and the diagnosis of infections of the urinary tract, AMA., *Arch. Intern. Med.*, 100, 709, 1957.
11. **Curby, W. A., Lind, H. E., and Moy, E. C.**,Dynamic state measurements of the degree of microbial resistance to antibiotics, presented at the 71st Annu. Meet. Am. Soc. Microbiol., Minneapolis, Minn., May 1971, 64.

Chapter 6

INSTRUMENTS ON OR NEAR MARKET

I. INTRODUCTION

There are several different approaches to microbiological analysis of body fluids which have been applied to sophisticated instrumented systems performing one or more parts of this analysis. Among the principles of action used in instrumented systems either on or approaching the market are radiometric assay, photometry, and electrical impedance. Each of these principles will be illustrated by discussing six instrumented systems separately with respect to the name of the manufacturer, their principle of action, the application of this principle to the analysis, a description of the system, methods for each analytical application, instrument features and requirements, market status, general comments, practical considerations, and references (prices, May 1978).

II. RADIOMETRIC APPROACH — BACTEC® INSTRUMENTED SYSTEM

Bactec® is the registered trademark for a series of instruments using automatic processes introduced in 1970 by Johnston Laboratories, Inc., Cockeysville, Md. 20130 which applies the radiometric principle to microbiological analysis of body fluids, principally to detecting bacteria in blood cultures and other normally sterile body fluids.

Bactec® is patented: U.S. 3,676,679; other U.S. and foreign patents granted or pending. All rights reserved.

A. Fundamental Principle of Action

Radiometric microbiology is based on the detection of radioactive gas produced by metabolism or degradation of radiolabeled substrates by microbial or biochemical action, applied to clinical microbiology by DeLand and Wagner.[1] Viable microorganisms present in a test sample inoculated into a closed culture vial containing growth medium and ^{14}C substrates will metabolize the radioactive material and produce $^{14}CO_2$ as one end product. The instrument analyzes the atmosphere in the culture vial for the presence of $^{14}CO_2$, indicating positive growth when the level of $^{14}C_2$ exceeds a threshold value.

B. Application of Principle

Microbiological assays measuring $^{14}CO_2$ produced from ^{14}C-labeled substrates, such as glucose or protein metabolites, are based on the fact that almost all bacteria will utilize these substrates and produce $^{14}CO_2$ as one end product.

The major application of the radiometric technique in clinical microbiology is the detection of bacteria in specimens of blood as well as in urine, spinal fluid, synovial, and pleural fluids. The radiometric procedure has high sensitivity and will give a positive signal when the culture has reached approximately $5 \times 10^5 - 10^6/m\ell$ bacteria which permits rapid detection of bacteria in a normally sterile clinical specimen. Enumeration of bacteria can be performed on the original sample by determining time-to-detection. Antibiotic susceptibility and identification tests can be performed on pure cultures, although the latter two tests are not now being done routinely, except for two special tests.

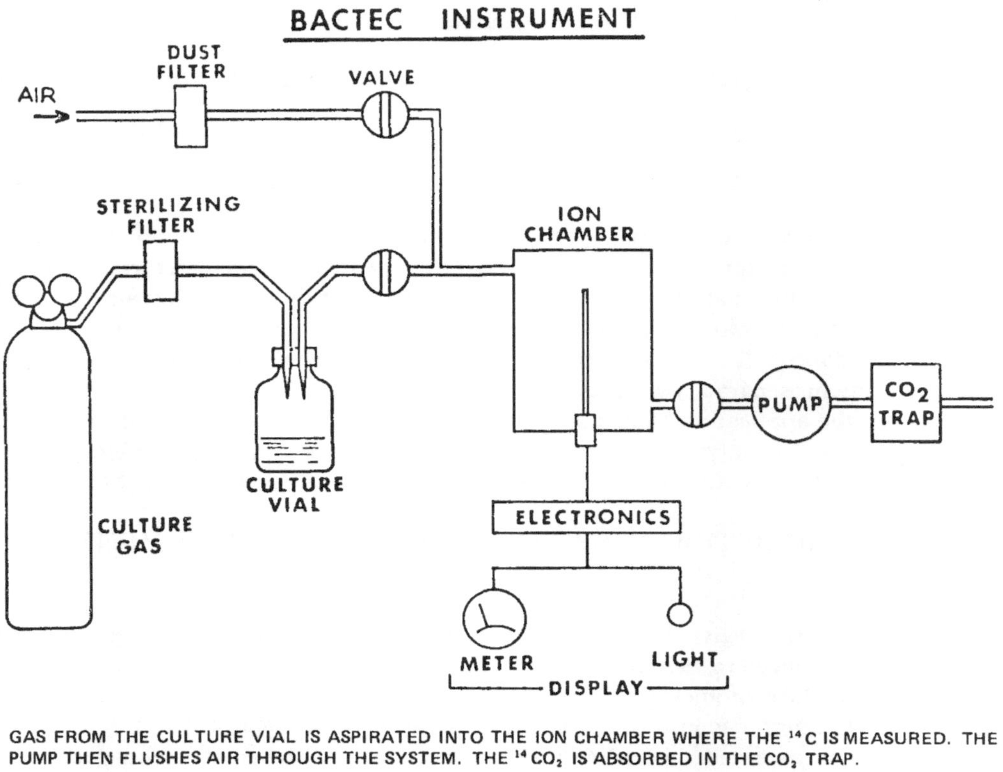

FIGURE 1. Schematic Diagram of Bactec® Instrument. (From Johnston Laboratories, Inc. With permission.)

C. Instrument Design and Operation

All three Bactec® instruments share a common operational scheme, best described with the aid of the simplified instrument schematic diagram in Figure 1 and pictured in Figure 2.

The essential components include the culture vial connected to the ion chamber, an electronic meter or printer to display results, pump, gas, sterilizing apparatus, and a CO_2 trap.[2, 3] The operation of instruments is automatic after the culture vial containing the properly prepared sample is placed in the instrument.

The operating procedure is simple and straightforward. A culture vial, sealed with a rubber septum, containing reagents and about 1.5 μCi radioactive substrates is inoculated and loaded into the instrument. Two sterile needles puncture the rubber septum and penetrate into the gas space below the septum, but above the solution in the vial. The instrument test cycle begins with room air being drawn in through the dust filter and through the ion chamber by the pump to purge the system. The flow of air is stopped, and a reference reading of background radioactivity in the chamber is made and stored electronically. The ion chamber is then partially evacuated by the pump. When the sample inlet valve is opened, culture gas is drawn from the external gas cylinder through a sterilizing filter, into the vial by one needle, out of the vial by the other needle, and into the ion chamber, sweeping out the atmosphere in the test vial and refilling the vial with sterile culture gas.

This determination is performed inside the chamber, as beta particles emitted by any $^{14}CO_2$ present in the sample gas will cause a small, but measurable, current to be

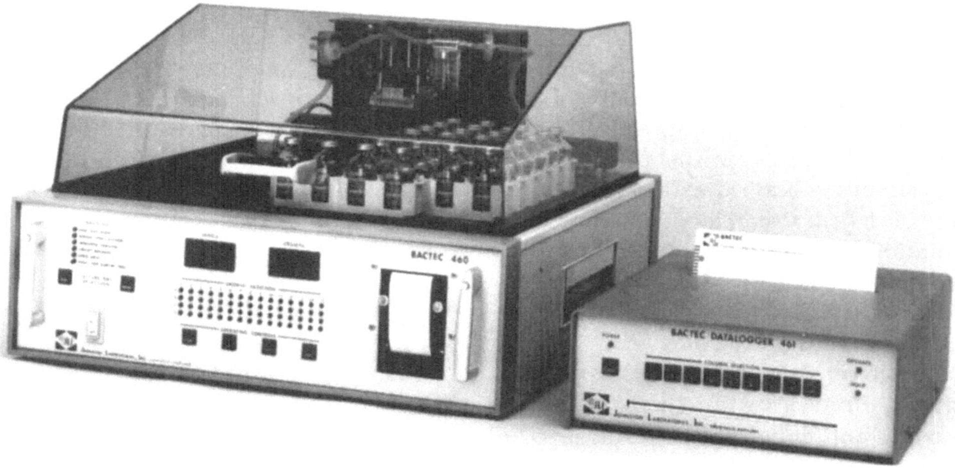

FIGURE 2. Bactec® Model 460 with Datalogger. (From Johnston Laboratories, Inc. With permission.)

generated at the chamber anode. The anode current is measured by an electrometer which automatically subtracts the previously obtained background reading. The resulting signal, now proportional only to the amount of radioactivity sampled from the culture vial, is conditioned by the remainder of the instrument electronics to provide meter or printer and display readout, designated as the Growth Index.

The needles are then withdrawn from the vial and the ion chamber again flushed with air to remove any residual radioactivity, thus preventing carry-over of radioactivity to the next test. The needles are then sterilized by heating to over 160°C and purged to insure that no microbial contamination is introduced into subsequent vials. Purge and sample gases exiting the instrument are passed through a CO_2 trap containing 1.2 kg soda lime, effectively preventing the release of any radioactive gas into the laboratory environment.

D. Methods

1. Detection

To detect the presence of bacteria, the specimen, such as blood, is placed in a closed vessel together with growth media containing about 1.5 μCi radiolabeled carbon substrates, a background reading is made, and the culture is incubated with stirring, usually in both an appropriate aerobic and anaerobic environment. The increase in $^{14}CO_2$ produced is measured at intervals, typically every few hours for several readings and with decreasing frequency thereafter. When a specified threshold, such as a scale reading of 20 or two times background, is reached the culture is considered positive for the presence of bacteria. The detection level is about 0.005 μCi!

2. Enumeration

It is obvious that bacteria can be detected in the original specimen, but what is not obvious is that this approach can be used to quantify the number of microorganisms in the specimen. Enumeration of the bacteria in the original sample can be determined within the framework of a specific protocol by the time required to reach the detection level, as it has been demonstrated experimentally by Waters[4] that the more bacteria present in the test sample, the more $^{14}CO_2$ is liberated and the quicker the detection level is reached. In fact, there is a linear relation betwen this time-to-detection and the reciprocal of the logarithm of the number of cells in the initial inoculum in the test vial. The parameters in this relationship vary depending on the type of organism, the cultural conditions, previous history, etc. Nevertheless, once a particular test protocol is defined and the test system calibrated, the time-to-detection can be used to quantify the number of organisms present in the original sample. For many organisms, faster detection is obtained as the volume of culture medium used is decreased. Of course, for blood culturing, a high volume of culture medium must be used to maintain a 10:1 dilution ratio for the blood. For other uses, however, it is usually best to restrict the volume of culture medium to minimum.

3. Antibiotic Susceptibility Tests

Antibiotic susceptibility tests of pure cultures can be performed by detecting growth/no growth or growth reduced by 50% when compared to a control in the presence of various antibiotics or at various levels of one antibiotic (minimum inhibitory concentration) (MIC).[5, 6] This is a rapid test, yielding results within 3 hr after inoculation with a pure culture. Johnston Laboratories also offers a kit for the serum assay of the aminoglycoside class of antibiotics.[7]

4. Identification

Using a pure culture, identification can be made in a like manner with a battery of selective or inhibitory media. This procedure is being used for identification of pathogenic *Neisseria* in pure culture[8] using the *Neisseria* Differentiation Kit.

Neither routine antibiotic susceptibility testing or identification, except as noted, is being claimed using present radiometric instrumentation.

E. Instrument Features

Johnston Laboratories currently manufactures three separate Bactec® instruments, each suited to a particular type of investigation and/or sample load. Johnston Laboratories' *Guide for the Selection of Your Bactec System* contains this information — the three Bactec® systems have the following general suitability:

1. Model 301: manual operation for the small hospital doing a limited number of blood cultures (20 to 60/week)
2. Model 460: for the hosptial wanting the advantages of automation (50 to 350 blood cultures/week)
3. Model 225: fully automated with built-in incubation and agitation (100 to 350 blood cultures/week)

Specific information on each model is as follows:

	Model		
Feature	301	460	225
Culture vial load per test cycle	1	60	25
Automated culture vial change	No	Yes	Yes
Automatic test cycle time	90 sec	1 hr	20 min
Automatic test cycle repeat	No	No	1-, 2-, 3-hr selector
Automatic sampling needle sterilization	Yes	Yes	Yes
Automatic vial top sterilization	No	No	Yes
Built-in incubation/agitation	No	No	Yes
Panel readout	Meter	Digital display	Digital display
Tape printout	No	Yes	Yes

		Model		
Feature		301	460	225
Positive sample lights		No	Yes	Yes
Accessories required:	Culture Gas Selector	Model GA-1	Built-in	Model GA-1
	Aerobic Gas Adaptor	P/N 225-026	P/N 460-020	P/N 225-026
	Anaerobic Gas Adaptor	P/N 225-025	P/N 460-019	P/N 225-025
	Shaker or shaker/incubator	not supplied; to be purchased separately	not supplied; to be purchased separately	Built-in
Optional:	Data card recorder	No	Model 461 Datalogger	No

F. Instrument Requirements

1. All Bactec® instruments require the use of Bactec® culture media which include ^{14}C substrates.
2. A means must be provided for the incubation and agitation of aerobic clinical cultures when using the Model 460 or the Model 301. Continuous agitation and incubation both are provided in the fully automated Model 225
3. Anaerobic culturing with the Bactec® instruments requires the use of an anaerobic culture gas, usually nitrogen containing 5% H_2 and 10% CO_2. Aerobic culturing may be accomplished using air as the culture gas, but compressed air containing 5% CO_2 is recommended to enhance the growth of some organisms. Culture-gas adaptors are supplied as accessories for the Models 460 and 301, but are furnished with the Model 225, including a culture gas selector. A culture-gas selector is an integral part of the Model 460, but must be purchased as an accessory for the Model 301.

G. Market Status[9]

The basic Bactec® system, comprising the Bactec® instruments and clinical media, is fully developed. The first-production Bactec®, a Model 225, was sold and placed in service in the fall of 1970. Since that time, the Bactec® instrument line has been expanded to include Models 301 and 460. Test kits for in vitro diagnostic tests of serum aminoglycoside levels and for the differentiation of the pathogenic *Neisseria* have also been added to extend the usefulness of the instruments. New instruments, new or improved media, other in vitro diagnostic tests, and the like continue to be developed to augment the present product line. Market acceptance of the Bactec® system is good. User experience is always the prime determinant of final acceptance in the critical and hence conservative market for clinical laboratory instrumentation. Clinical users of the system now number over 500 hospitals, many with over 5 yr of clinical experience.

H. General Comments[9]

1. Numerous papers favorably comparing Bactec® with traditional blood culturing methods have been published, including several large studies.[10-12]
2. Recently, the use of radiometric procedures to detect the growth of *Mycobacterium tuberculosis* have been reported by Middlebrook et al.[13] and Truant et al.[14] and to test antimicrobial susceptibility by Foley et al.[15] A 1-hr screening for bacteruiria was described by Frank et al.[16]
3. As one of the first instrumented systems to be introduced to the microbiologist, Bactec® has been an important contributor of automation to a field previously dedicated to manual operations. Acceptance continues to grow as the advantages of reliable automated processes continue to be realized by the clinical, industrial, and research microbiological communities.

4. The Bactec® radiometric method properly utilized features rapid and predominantly objective reporting of results in a variety of applications. Little or no sample preparation is required in most instances. Samples, substrates, or reagents need not have restrictive optical, physical, or chemical properties. Naturally, substances interfering with the specific of a given test must be removed or their effects minimized by appropriate dilution. The high sensitivity of the radiometric method lends itself well to situations normally requiring gross visual or chemical changes to indicate test results. Additionally, the ability to use the system with modified radiotracer techniques in research permits the investigation of metabolic pathways and chemical or biochemical reaction mechanisms provided such pathways or mechanisms produce a radiolabeled, beta-emitting gas.
5. The radiometric principle can also be applied to bacterial enumeration in the production of foods and in-house sterility testing of foods, pharmaceuticals, drugs, radioisotopes, and other products.
6. Clinically, proper agitation of aerobic cultures is necessary. The interpretation of instrumented results is optimized when done by persons well acquainted with the system by virtue of training and experience. Used properly within the context of a reliable test protocol, the Bactec® system provides rapid detection of positive cultures, with few false positives or false negatives. Because no bacterial detection system can be considered perfect, a terminal blind subculture of all negative vials is recommended.

I. Practical Considerations

The principle of action of the radiometric procedure as it is applied to microbiological analysis is straightforward; the instrumentation is simple. However, there are several practical considerations in implementing the technique that deserve discussion. These include types and composition of media, collection of the gas, and laboratory and legal aspects of handling radioactive material.

1. Media and Culturing Environment

a. Radiometric analysis is conducted on clinical specimens, especially blood, that may carry very low levels of fastidious or damaged bacteria that have critical growth requirements, both with respect to media, nutritive composition, and tonicity and gaseous atmosphere. A rich broth must be selected to provide the nutrients required for the rapid growth of the fastidious bacteria. In addition, the radiolabeled substrates chosen must provide a source of ^{14}C for the production of $^{14}CO_2$ by the expected pathogens in the various samples. Glucose provides a good starting material which needs augmentation with protein metabolites to cater to nonfermenters.
b. Many clinical specimens are collected from patients receiving antibiotics that damage the cell wall of susceptible bacteria, leaving the organism vulnerable to the usual hypo- or even iso-) tonic media. For this reason, hypertonic media has been devised and has proven superior for the isolation of bacteria from those patients on certain antibiotics.[17]
c. There are cases in which knowledge about a specific organism is desired. By coupling selective substrate additives with a selective medium, the growth rate of the desired organism can be enhanced. Interfering organisms may grow slowly under the less stringent conditions, but are not detected, since they produce no radiolabeled gas from the selective substrate. The result is an organism-specific test, providing quantitative information far more rapidly than possible using conventional techniques.[9]
d. It is possible under favorable conditions to choose selective substrates in order

to adjust the times-to-detection of various organisms in mixed culture so as to provide indication of a given number of organisms singly or in combination after a specified incubation period. In this case, specific substrate utilization rates are used to compensate for differing bacterial metabolic rates. Such a test can provide mixed flora total count information much more quickly than the typical 48-hr plate count.

e. Bacteria present in clinical specimens may be aerobic, microaerophilic, or anaerobic. The medium environmental conditions in the culture vial must match their needs. The aerobic environment is easy to produce, but microaerophils do better in a less aerobic environment and are often detected more easily in a 10% CO_2 atmosphere. To detect the more strict anaerobes, prereduced media plus an anaerobic atmosphere of 85% N_2, 10% CO_2, and 5% H_2 give better results.

f. CO_2 is soluble in culture media, necessitating the use of some method of releasing the $^{14}CO_2$ so that it can be measured. This can be accomplished by lowering the pH below 5, which creates undesirable conditions for growth, or by reducing the atmospheric pressure, which is not practical. An alternate method is mechanical agitation, such as continuous stirring of the culture with a magnetic stirring bar which forces the $^{14}CO_2$ out of solution. This is the method used in the Bactec® instrument systems. Certain media containing bulky constituents, such as gelatin, hold the CO_2 more strongly and should be avoided.[18]

g. Red blood cells can metabolize glucose to form CO_2, and although this activity is at a low level, it does create a background. Many media lyse the blood cells which eliminates this reaction; these media should be selected for use with blood cultures.[18]

2. *Method of Measurement*

Radioactive CO_2 can be measured in several devices, such as a liquid scintillation counter or as a free gas in an ionization chamber in conjunction with an electrometer. Although the latter method is somewhat less sensitive, it is adequate for the purpose needed, and since it is much easier to automate, it has been chosen for measurement in the Bactec® instrument system.

3. *Radioactive Material Handling*

The handling of radioactive material requires some caution, both with respect to laboratory safety and legal requirements. $^{14}CO_2$ emits only low-energy electrons which are stopped by a glass vial and rubber seal. It is principally a safety hazard only if inhaled or ingested. Thus, precautions must be taken to prevent escape of the radioactive gas into the atmosphere, which can be done by installing a CO_2-absorbing trap, such as soda lime CO_2 trap at the exit of the chamber.

Government regulations to possess, use, or dispose of radioactive materials do not extend to the small quantities of radioactive material required for this test; however, the institution's radiation safety officer should be consulted for local restrictions as advised by Bactec® advertising literature.

J. Other Instruments

Another instrument using $Ba(OH)_2$ pads to collect the $^{14}CO_2$ produced by pure cultures of bacteria growing in selective media which are read in a Geiger-Muller gas-flow counter has been studied for bacterial identification by Schrot et al.[19] and patented by Biospherics, Inc., Rockville, Md. (U.S. Patent 4,057,470, 1977).

Much of the information in this chapter is summarized briefly in Tables 1 to 3.

TABLE 1

Instrument Designation, Principle of Action and Applications of Bactec®

Instrument tradename	Sponsoring company	Principle of action	Principle of Action to: (XX = principal application; X = secondary or potential use)				
			Detection	Enumeration	Antibiotic susceptibility	MIC[a]	Identification
Bactec® (all models)	Johnston Laboratories, Inc., Cockeysville, Md.	Radiometric; $^{14}CO_2$ released from labeled substrates by bacterial metabolism	XX Release of $^{14}CO_2$ from labeled substrates (mainly used for blood and other normally sterile body fluids)	X Time to detection level of released $^{14}CO_2$	X Serum assay kits for aminoglycoside antibiotics by inhibition of $^{14}CO_2$ released from ^{14}C-urea		X (Selective substrate utilization) Neisseria Differentiation Kits by $^{14}CO_2$ release from labeled differential sugars

[a] Minimum inhibitory concentration.

TABLE 2

Requirements and Features of Bactec®

	Requirements			Features					
Instrument tradename	Pure culture	Special media	Other	Degree of automation	Test period	Number of tests per day per unit	Accuracy compared to conventional tests	Strengths	Weaknesses
Bactec® Model 225	For identification	Yes	Requirements for special precautions because of radioactive material	Fully automated vial processing, incubation, agitation, repeat cycle at 1-, 2-, 3-hr intervals	Several hours to several days (blood), 1—3 hr (urine)	Up to 600	Good	Rapid detection of viable organisms	Requires use of radiolabeled ^{14}C substrates
Bactec® Model 460		Yes		Fully automated vial processing, but without incubation, agitation		Up to 500	Good		
Bactec® Model 301		Yes		Partially automated		Up to 280	Good		

TABLE 3

Market Aspects of Bactec®

Instrument tradename	Clinical trials	FDA status	Instrument price		Consumables cost		Commercial potential
			Sale ($)	Lease/rent ($)	Per vial ($)	Per test ($)	
Bactec® Model 225	Many	Registered	35,000 (total system)	Available for all models[a]	1.06—1.25 (6B, 7B, 8A media) (1 anaerobic, 1 aerobic, vial/test)	2.12—2.50	In addition to clinical uses, can be applied to enumeration, sterility tests on foods, pharmaceutical drugs, radioisotopes, etc.
Bactec® Model 460		Registered	12,401 (instrument) 12,900 (basic system) 15,150 (system with 461 Datalogger)		1.06—1.25	2.12—2.50	
Bactec® Model 301		Registered	3,445 (instrument) 4,284 (complete system)		1.06—1.25	2.12—2.50	
ND-1®, Neisseria Differentiation Kit		Registered	29.50 (10 tests)			2.95	—
SAG-1®, SAT-1®, Serum Aminoglycoside Assay Kits Gentamicin, Tobramycin		Registered	120 16—70 tests)			1.70—7.50	—

[a] For terms see domestic price list S-2A.

REFERENCES

1. **DeLand, F. H. and Wagner, H. N., Jr.**, Early detection of bacterial growth with carbon-14-labeled glucose, *Radiology,* 92, 154, 1969.
2. **Waters, J. R. and Zwarun, A. A.**, Results of an automated radiometric sterility test as applied to clinical blood cultures, in *Developments in Industrial Microbiology,* Vol. 14, Murray, E. D., Ed., Hennage Creative Printers, Washington, D.C., 1973, chap. 9.
3. **Morello, J. A.**, Automated radiometric detection of bacteremia, in *Automation in Microbiology and Immunology,* Hedén, C. G. and Illéni, T., Eds., John Wiley & Sons, N.Y., 1975, chap. 3.
4. **Waters, J. R.**, Sensitivity of the $^{14}CO_2$ radiometric method for bacterial detection, *Appl. Microbiol.,* 23, 198, 1972.
5. **DeBlanc, H. J., Jr., Charache, P., and Wagner, H. N., Jr.**, Automatic radiometric measurement of antibiotic effect on bacterial growth, *Antimicrob. Agents Chemother.,* 2, 360, 1972.
6. **DeLand, F. H.**, Metabolic inhibition as an index of bacterial susceptibility to drugs, *Antimicrob. Agents Chemother.,*2, 405, 1972.
7. **Buda, D. and Broman, R.**,Radiometric Assay of gentamicin in blood by measurement of bacterial evolution of $^{14}CO_2$ from ^{14}C-urea, presented at the 75th Annu. Meet., of the Soc. Am. Microbiology, New York, April-May 1975, 31.
8. **Foley, T. and Broman, R.**, Radiometric identification of pathogenic *Neisseria* by $^{14}CO_2$ production from labeled sugars, presented at 75th Annu. Meet. Soc. Am. Microbiology, New York, April-May, 1975, 31.
9. **Ahnell, J. E.**,personal communication, 1978.
10. **DeBlanc, H. J., Jr., Deland, F., and Wagner, H. N., Jr.**, Automated radiometric detection of bacteria in 2967 blood cultures, *Applied Microbiol.,* 22, 846, 1971.
11. **Randall, E. L.**, Long-term evaluation of a system for radiometric detection of bacteremia, in *Microbiology 1975,* Schlessinger, D., Ed., American Society Microbiology, Washington, D.C., 1975, 39.
12. **Jacobs, P. F., Bannister, E. R., Burdash, N. M., and West, M. E.**,Evaluation of the BACTEC 460 in a large clinical laboratory, presented at 77th Annu. Meet. Am. Soc. Microbiology, New Orleans, May 1977, 65.
13. **Middlebrook, G., Reggiardo Z., and Tigertt, W. D.**, Automatable radiometric detection of growth of *Mycobacterium tuberculosis* in selective media, *Am. Rev. Respir. Dis.,* 115, 1066, 1977.
14. **Truant, J., Muller, B., Frazier, G., and Stroup, G.**,Radiometric detection of *Mycobacterium,* presented at the 78th Annu. Meet. Soc. Am. Microbiology, Las Vegas, May 1978, 301.
15. **Foley, T., Truant, J:, Muller, B., and Zeig, D.**,A radiometric procedure for testing the susceptibility of *Mycobacterium,* presented at 78th Annu. Meet. Am. Soc. Microbiology, Las Vegas, May 1978, 296.
16. **Frank, H. G., Wiener, L. B., and Waters, J. R.**, Evaluation of a one-hour radiometric screening test for urine cultures presented at 78th Annu. Meet. Am. Soc. Microbiology, Las Vegas, May 1978, 301.
17. **Coleman, R. M., Laslie, W. W., and Lambe, D. W., Jr.**, Clinical comparison of aerobic, hypertonic and anaerobic culture media for a radiometric detection of bacteremia, *J. Clin. Microbiol.,* 3, 281, 1976.
18. **Gall, L. S., Plaugher, R. N., Stiner, M. S., and Tulis, J. J.**, Studies on a radiometric method for rapid detection of bacteremia, including criteria for media selection, in *Developments in Industrial Microbiology,* Vol. 13, Murray, E. D., Ed., Garamond/Pridemark Press, Baltimore, 1972, chap. 34.
19. **Schrot, J. R., Hess, W. C., and Levin, G. V.**, A method for the rapid radiorespirometric identification of bacteria, presented at the 74th Annu. Meet. Am. Soc. Microbiology, Chicago, May 1974, 113.

Chapter 7

PHOTOMETRIC APPROACH TO MICROBIOLOGICAL ANALYSIS

I. INTRODUCTION

Several instrumented systems designed for microbiological analysis, especially antibiotic susceptibility testing, use photometric techniques as the method of measurement of changes in bacterial populations. The fundamental principle of action of photometry has been previously discussed in the chapter on theoretical aspects. The various photometric approaches used in instrumented systems include turbidity, forward light scattering at a single angle, and differential light scattering. All of these methods measure the effect of the bacterial culture and the experimental conditions imposed on that culture on a light beam by determining the intensity of light passing through the culture. This is measured with a type of photoelectric cell set at some specific angle to the axis of the incoming light beam.

Several instrumented systems apply sensitometry or densitometry to microbiological analysis. These include Pfizer's Autobac 1®, Science Spectrum's Differential III®, McDonnell-Douglas's AMS®, and Abbott's MS-2®. Although these techniques appear to be similar, there are some important differences in the measurements which may affect their suitability for use in certain aspects of microbiological analysis, and the specific principle of action of each will be discussed separately in the appropriate chapter.

Three of these systems are designed primarily for antibiotic susceptibility testing, but there is also a potential with most of these systems for using photometry for detections and identification using pure cultures.

II. AUTOBAC 1®

Autobac 1® is the registered trademark of an instrumented system, using automated processes for antibiotic susceptibility testing and produced by Pfizer Diagnostics Division, New York, N.Y. This approach measures light emerging from a cuvette cell containing a bacterial culture to detect changes in bacterial populations in that sample. Pfizer holds U.S. Patent 3,832,532, Design Patent 230,587, copyright 1974, all rights reserved.

A. Fundamental Principle of Action

Light dispersion can be used in microbiological analyses because it can be shown that the optical density of a sampling vessel containing growth media and growing bacteria will increase as the bacterial population increases.[1,2]

The Autobac 1® shines white unpolarized light through the broth culture and measures the light leaving the sample at a fixed angle of $35 \pm 2°$ from the axis of the incoming light beam by means of a photosensor. The angle chosen was determined by an attempt to obtain the greatest light intensity (sensitivity) compatible with retaining the greatest change in optical density per change in bacterial population number.

The amount of light emerging from a standardized bacterial suspension in the presence and absence of antimicrobial agents is read by a photometer at a single angle of measurement and is converted into a number referred to as a light scattering index (LSI). This is related to the percentage inhibition of the test culture as compared to the log of the growth in the control chamber.

Methods of measuring light dispersion at only one angle can detect changes in cell mass as brought about by an increase either in cell population or cell size, but cannot distinguish which. However, forward-angle light measurement offers a marked increase in sensitivity over turbidimetric systems which read the reduction of the direct beam of light passing through the culture.

B. Application of Principle

The principle application claimed for Autobac 1® at present is for antibiotic susceptibility testing, although there is a potential for detection and enumeration of bacteria in the original specimen and the identification of pure cultures of bacteria.

1. Antibiotic Susceptibility Testing

An antibiotic susceptibility test measures the effect of various antibiotics on the biomass of a bacterial culture compared to that of a control. Since photometric techniques can measure biomass, the approach used in Autobac 1® is suitable for antibiotic susceptibility tests by determining the degree of increase or lack of increase in biomass in an antibiotic-containing test culture when compared to the control. Lack of increase in cell mass is interpreted as indicating susceptibility of the culture to the test antibiotic, while cell mass similar to the control indicates resistance. In-between values at a specified level are designated as intermediate.

Using a variation in this technique, the minimum inhibitory concentration (MIC) of an antibiotic can be determined by challenging the test culture with more than one concentration of an antibiotic and measuring the effect of each concentration on the biomass. The highest dilution of antibiotic which inhibits an increase in cell mass represents the MIC of that antibiotic for that bacterium. The Autobac 1® method employs calculations based on data derived from two concentrations of the antibiotic to arrive at the MIC value.

2. Detection and Enumeration

The detection of viable bacteria in a clinical specimen is based on an increase in biomass during an incubation period.[1,2] Since forward-angle light dispersion techniques can measure an increase in biomass, this approach can be used for detection of bacteria. Enumeration can also potentially be performed with this instrument system as the emerging light measured has been shown empirically to be proportional to the bacterial population over a three log growth range.

3. Identification

Bacteria are often identified by characteristic patterns shown by reactions to selective or inhibitory media. Since light dispersion techniques can detect growth/no growth by measuring biomass, characteristic patterns based on growth/no growth may be determined and aid in the identification of pure cultures of bacteria.

C. Instrument Design and Operation[1,2,3]

The Autobac 1® system is a semiautomated system for antimicrobial susceptibility testing in a clinical laboratory that can perform susceptibility tests on 12 antimicrobial substances using pure cultures within 3 to 5 hr. Each antibiotic receives a numerical value of 0 to 1 which is interpreted by the system as resistant, intermediate, or sensitive. A report ticket is then printed with the numerical result and the interpretation.

The Autobac 1® system (Figure 1) consists of four basic components.

1. Cuvette

The cuvette (Figure 2) is a disposable, optically clear 13-chambered plastic device

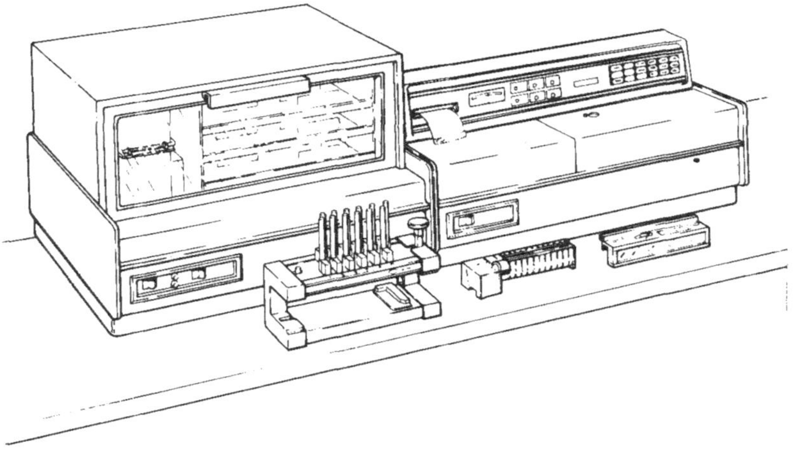

FIGURE 1. Sketch of Autobac 1® system. (From Pfizer Diagnostics. With permission.)

with a broth inoculum chamber, which contains a built-in means for distributing the Eugonic broth inoculum in equal amounts to 12 test chambers and a control chamber. The inoculum is previously standardized by a built-in photosensor.

2. Disc Dispenser

A disc dispenser loads the cuvette with a panel of 12 antimicrobial elution discs.

3. Incubator/Shaker

An incubator/shaker agitates the cuvette contents employing a ¾ in. amplitude rotary motion at 220 rpm and maintained 36 ± 0.5°C.

4. Photometer

A photometer automatically measures through a 4° arc of the light emerging at 35° from the light beam axis from a quartz halogen lamp by a collimated photodetector and reads each chamber of the transparent cuvette for reduction in light caused by the bacterial suspension. The results are printed on a reporting ticket.

Autobac 1® performs the antibiotic susceptibility test automatically after the antibiotic discs are added, the standardized inoculum dispensed, and the cuvette placed in the instrument.

D. Methods[1,2,3]

1. Antibiotic Susceptibility Tests

To perform an antibiotic susceptibility test using Autobac 1®, a specimen is plated, and after incubation, morphologically identical colonies are picked and suspended in buffered saline solution. The concentration of the inoculum is adusted in the instrument photometrically to approximately 1.5 to 3.0 × 10^7 organisms per milliliter and diluted in Eugonic broth to 1.5 to 3.0 × 10^6 organisms per milliliter, which photometric value (G_k) is used in the calculation of LSI. The inoculated broth tube is then screwed into the cuvette containing the susceptibility elution discs, and the assembly is manipulated and positioned so that equal amounts of broth are in each measuring chamber after which the cuvette is incubated at 36°C for at least 3 hr with shaking to properly suspend the organism and elute the antibiotic.

After 3 hr, the cuvette is read automatically by the Autobac 1® photometer, and a

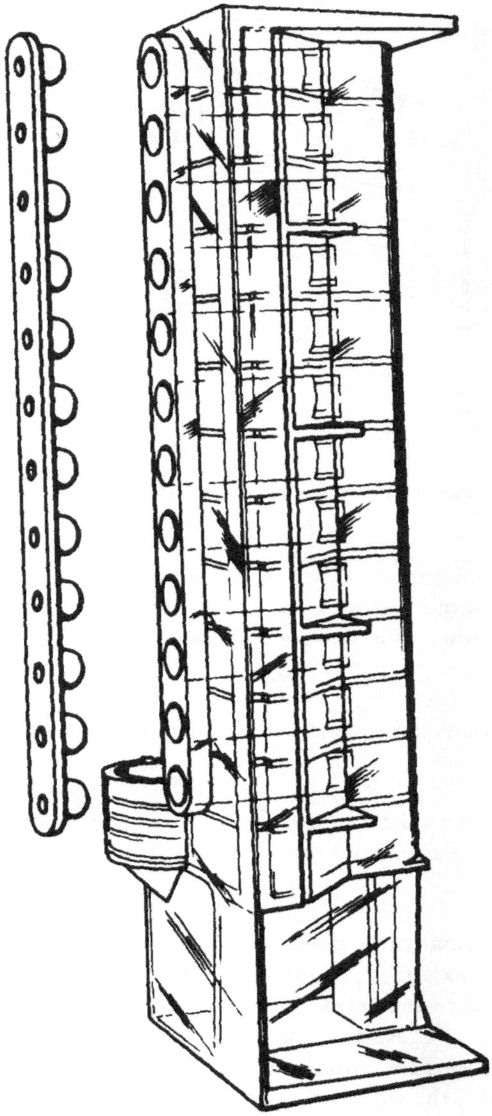

FIGURE 2. Sketch of cuvette of Autobac 1® system. (From Pfizer Diagnostics. With permission.)

calculation of LSI for each antibiotic is made if there has been sufficient growth. The logarithm of bacterial concentration (G_c) in the control chamber of the cuvette (no antimicrobic) is compared with the logarithm of bacterial concentration (G_x) in each of the 12 test chambers with the antimicrobic. The LSI calculated based on G_c, G_x, and G_k is a measure of the inhibition of growth in each chamber on a scale of 0.00 to 1.00. The interpretation of the data supplied by the photometer for all antimicrobials except penicillin G is (from package circular):

Interpretation	Symbol	LSI Range
Resistant	R	0.00—0.50
Intermediate (equivocal)	I	0.51—0.59
Susceptible	S	0.60—1.00

Penicillin G is interpreted as:

Interpretation	Symbol	LSI range
Resistant	R	0.00—0.90
Susceptible	S	0.91—1.00

An antibiotic test determining the MIC is being studied using Autobac 1®. Brown and Sautter[4] and Gavan et al.[5] have adapted this system to perform an MIC test using two concentrations of each antibiotic rather than a single concentration of various antibiotics.

2. Detection, Enumeration, and Identification

The instrument system has potential to detect and enumerate bacteria, since the photodetector output voltage is directly proportional to the number of organisms and may be used as a sensitive index of bacterial population by measuring biomass.[1] Identification of bacteria may be performed by establishing a profile derived from LSI values using differentially inhibitory agents. Sielaff et al.[6] have studied this application, and recent experiences using Autobac 1® in the identification of Gram-negative bacteria and Gram-positive bacteria were presented at a company-sponsored user symposium on new applications of the Autobac 1® system.

E. Instrument Features

The Autobac 1® offers a semiautomated system with several unique design features.

1. The photometer measures emerging light at a 35° angle to the incident beam, which was selected as the best compromise between sensitivity and instrument simplicity.
2. A complex disposable cuvette of optically clear polystyrene has 13 test chambers and contains a built-in means for distributing the broth inoculum in equal amounts of 1.5 m*l* to each chamber and placing the inoculum in contact with the antimicrobial-containing discs.
3. The disc dispenser is specially constructed to place the 12 discs simultaneously into the chamber of the cuvettes.
4. The incubator/shaker is designed to hold cuvettes and is an integral part of the instrument.
5. After the cuvettes are placed in the incubator, the procedure is automatic except for manual transfer of the cuvettes from incubator to photometer. Results are obtained in 3 to 5 hr.
6. Pfizer offers an electronic data processing package that links the results from Autobac 1® with a central computer to produce antibiotic susceptibility patient profile data.

F. Instrument Requirements[3]

1. Autobac 1® requires the use of the Autobac 1® line of expendables, including

cuvettes, Eugonic broth, inoculum standardization solution and tubes, reporting forms, and antimicrobial-susceptibility elution test discs.

According to their product literature, the accuracy, reproducibility, and reliability of the system are contigent upon the use of Autobac 1® expendables. System performance cannot be guaranteed if expendables other than those supplied by Pfizer Diagnostics are used.[3]

2. Also required and supplied with the instrument are the Autobac 1® photometer, test/calibration wedge, and incubator indicators.
3. Required, but not supplied, are ordinary laboratory culturing equipment and specified control organisms.
4. The antibiotic susceptibility test must be performed on pure cultures.
5. Since the system requires the picking of multiple colonies, there is an element of risk that two "similar" colonies may indeed be different organisms.

G. Market Status

Autobac 1® was introduced in 1975, and over 300 units have been sold or leased to various hospital laboratories. Autobac 1® system has been cleared by the Food and Drug Administration (FDA) for single elution disc for antibiotic susceptibility testing and MIC.

Clinical trials are underway to evaluate the instrument system to perform identification tests of pure cultures using selective agents.

H. General Comments

1. Autobac 1® used for antibiotic susceptibility testing offers rapid results in 3 to 5 hr after obtaining the pure culture.
2. The Autobac 1® system features, in addition, a high degree of reliability, simplicity of use, significant labor savings, and a markedly higher day-to-day reproducibility and accuracy than are attainable with manual techniques as they are routinely employed in the laboratory according to the product literature.
3. The speed, accuracy, reproducibility, and objectivity for the majority of antimicrobic organism combinations of Autobac 1® for antibiotic susceptibility testing have been confirmed in a major collaborative study coordinated by the Center for Disease Control.[7] The study was for 12 months, conducted in 7 laboratories, involving 26 antimicrobics against 2100 isolates in more than 22,000 tests. The method used in these studies proved to be easy to learn and perform, and the systems functioned well during these tests.
4. Other more recently published studies showed similar results.[8,9] Stubbs and Wicher[10] in a report involving 2518 freshly isolated cultures tested on Autobac 1®, found favorable overall agreement compared to Kirby-Bauer results, but they found discrepancies with certain organism/antibiotic combinations. This type of discrepancy is noted in the Autobac 1® package circular under Precautions and Limitations regarding certain organism/antibiotic combinations. In a personal communication, Frankel[11] of Pfizer Diagnostics states that certain of these discrepancies have since been overcome by alterations in disc concentration of antibiotic and improvement in handling of the disc during manufacturing and shipping of certain discs, such as ampicillin, and he further points out that the number of bacteria involved in the discrepancies in the Stubbs report is small.
5. Preliminary studies by Hale and Matson[12] indicated that Autobac 1® has potential for use as a 4-hr screening test for bacteriuria.

I. Practical Considerations

In designing the instrumented system and developing a method of operation, several practical aspects were considered.

1. Design Features[1,2]

a. Angle of Light

When photometric techniques measuring light at a single angle are employed, the angle chosen is of importance. Decreases in light intensity due to growth are much more pronounced at low angles. Shallower viewing angles offer greater sensitivity, but also require more complex optical systems. The choice of angle must be a compromise between sensitivity and simplicity. The choice depends on obtaining the highest average intensity of the light while maintaining the maximum differential of intensity with changes in concentration of biomass.

b. Wavelength

The instrument group developing Autobac 1® with fixed-angle photometry used unfiltered, unpolarized light from the tungsten halide source, since it is the most suitable for maximizing the sensitivity of the photometer for detecting changes in bacterial concentration in broth.

c. Cuvettes

As in most instrumented systems using photometry techniques, the cuvettes for Autobac 1® are fabricated especially for that instrument system. The Autobac 1® cuvette has 13 chambers, one for the control and the others for the test culture. The measurements from each of the test chambers are compared with the control. Therefore, the quality control for the cuvettes must ensure that the light beam passing through each chamber is affected in exactly the same way. Autobac 1® employs a distinctive cuvette that meets all the requirements of the system.

2. Operational Features

a. Incubation Time[1]

The incubation time for the test using Autobac 1® is based on the time needed for the antibiotic to exert a measureable effect on the test culture. Experiments conducted with Autobac 1® indicated that the test incubation period should be long enough for at least a three-generation bacterial growth, provided that this incubation period was not less than 3 hr. Earlier readings with certain bacteria/antimicrobial combinations, such as certain B-lactam-susceptible organisms, could produce false resistant results. Based on these data an incubation period of 3 to 5 hr was chosen.

b. Microorganism Concentration[1]

The microorganism concentration of the inoculum for most commonly isolated pathogens is chosen to give the maximum light intensity change during the test period. Studies with Autobac 1® showed that this occurred at 3 to 5 hr the chosen incubation time, when 10^6 bacteria per milliliter were present in the inoculum.

c. Reading

Before reading, inspect the cuvettes for bubbles that may reduce the light path through the fluid, giving a high reading, and for settling out of the bacteria which may give a low reading.

d. Scheduling

Autobac 1® determinations fit conveniently into a normal weekday working schedule, with the tests being set-up, read, and in report form by the end of the work day. However, in practice this may not be true of weekend and holiday schedules in some hospitals, and this aspect should be considered before acquisition.

Much of the information in this chapter is summarized briefly in Tables 1—3.

TABLE 1

Instrument Designation, Principle of Action and Applications of Autobac 1®

			Principle of action applied to: (XX = principal; application; X = secondary or potential use)				
Instrument tradename	Sponsoring company	Principle of action	Detection	Enumeration	Antibiotic susceptibility	Minimum inhibitory centration	Identification
Autobac® AIS (susceptibility) Autobac® MTS (MIC)	**Pfizer Diagnositics Division, New York, N.Y.**	Changes in light transmission at one angle in presence of bacteria	X Increase in biomass during incubation shown by decrease in transmitted light	X Decrease in transmitted light related to increase in biomass	XX Decrease or lack of decrease in transmitted light in presence of various antibiotics	XX As in previous column in presence of various concentrations of one antibiotic	X Characteristic pattern of growth/no growth in selective inhibitory media

TABLE 2

Requirements and Features of Autobac 1®

	Requirements			Features					
Instrument tradename	Pure culture	Special media	Other	Degree of automation	Test period	Number of tests per day per unit	Accuracy compared to conventional tests	Strengths	Weaknesses
Autobac AIS® (susceptibility)	Yes	Yes	Autobac line of expendables, such as cuvette, standardization equipment, antibiotic test elution discs	Automated after inoculum standardization and inoculation	3—5 hr	30 +	Good for most antibiotic/organism combinations	Rapid, accurate, reliable method	Less accurate determination of certain combinations of antibiotic/organism susceptibility as noted on package circular, method uses multicolony inoculum, inoculum not preincubated to attain log growth prior to antibiotic challenge

TABLE 3

Market Apects of Autobac 1®

Instrument tradename	Clinical trials	FDA status	Instrument price		Consumables cost		Commercial potential
			Sale ($)	Lease/rent	Per vial ($)	Per test ($)	
Autobac 1®	2100 cultures in seven hospitals	Elution discs approved by FDA for susceptibility and MIC testing	25,000	Yes	1.06—1.43	1.06—1.43	In addition to antibiotic susceptibilities, MIC, detection, enumeration, and identification with selective agents are under review and development

REFERENCES

1. **McKie, J. E., Jr., Borovoy, R. J., Dooley, J. F., Evanega, G. R., Mendoza, G., Meyer, F., Moody, M., Packer, D. E., Praglin, J., and Smith, H.,** Autobac 1 — a 3-hour automated antimicrobial susceptibility system. II. Microbiological studies, in *Automated Microbiology and Immunology,* Heden, C. G. and Illeni, J., Eds., John Wiley & Sons, New York, 1975, chap. 16.
2. **Praglin, J., Curtiss, A. C., Longhenry, D. K., and McKie, J. E., Jr.,** Autobac 1 — a 3-hour, automated antimicrobial susceptibility system. I. System description, in *Automated Microbiology and Immunology,* Héden, C. G. and Illéni, J., Eds., John Wiley & Sons, N.Y., 1975, chap. 15.
3. *The Autobac 1 System Operator's Manual,* May, 1977.
4. **Brown, W. J. and Sautter, R. L.,** Five-hour minimal inhibitory concentration test of four antimicrobial agents for *Pseudomonas aeruginosa, Antimicrob. Agents Chemother.* 11, 1064, 1977.
5. **Gavan, T. L., Baker, C. N., Schoenknecht, F. D., and Yu, P. K. W.,** Rapid determination of antimicrobial minimum inhibitory concentration using a modified Autobac 1. presented at 78th Annu. Meet. Am. Soc. Microbiology, Las Vegas, May 1978, 311.
6. **Sielaff, B. H., Buck, G. E., and Matsen, J. M.,** An innovative approach to bacterial identification using Autobac 1, presented at 77th Annu. Meet. Am. Soc. Microbiology, New Orleans, May 1977, 52.
7. **Thornsberry, C., Gavan, T. L., Sherris, J. C., Balows, A., Matsen, J. M., Sabath, L. D., Schoenknecht, F., Thrupp, L. D., and Washington, J. A.,** II, Laboratory evaluation of a rapid automated susceptibility testing system. Report of a collaborative study, *Antimicrobial. Agents. Chemother.,* 7, 466, 1975.
8. **Greenberg, J. M., Gottschall, R. L., and Randall, E. L.,** A comparison of Autobac 1 results with early and conventional readings of Bauer-Kirby disc diffusion tests, presented at 76th Annu. Meet. Am. Soc. Microbiology, Atlantic City, N.J., May 1976, 44.
9. **Phillips, S. B., Greenberg, D. S., Sierra, M. F., and Amsterdam, D.,** Autobac 1/Bauer-Kirby comparison and reproducibility: an interhospital evaluation of 604 clinical isolates, presented at 77th Annu. Meet. Am. Soc. Microbiol., New Orleans, May 1977, 40.
10. **Stubbs, K. G. and Wicher, K.,** Laboratory evaluations of an automated antimicrobial susceptibility system. *Am. J. Clin. Pathol.,* 68, 769, 1977.
11. **Frankel, H.,** personal communication.
12. **Hale, D. C. and Matsen, J. M.,** Clinical evaluation of screening urine specimens by light-scattering photometry, presented at 78th Annu. Meet. Am. Soc. Microbiology,, Las Vegas, May 1978, 301.

Chapter 8

DIFFERENTIAL III® INSTRUMENTED SYSTEM

I. INTRODUCTION

Differential® is a registered trademark for a series of photometers, (I, II, and III) measuring differential light scattering which are produced by Science Spectrum, Inc., Santa Barbara, California. Diff III®, a shortened form of Differential III®, is also a trademark of the firm. Issued U.S. Patents related to these systems include 3,730,842, 3,928,140, 3,770,351, 3,701,620, and 3, 624, 835.

Diff III® is intended to be used in the clinical laboratory for a variety of microbiology procedures, but particularly for antibiotic susceptibility testing.

II. FUNDAMENTAL PRINCIPLE OF ACTION

The principle of using light-scattering methods in microbiological analyses has been thoroughly discussed in a previous section on theory. The new Differential III® employs a 15-element detector array to measure monochromatic coherent vertically polarized light scattered as a function of angle. Such a measurement as differential light scattering[1,2] (DLS) allows the measurement of changes in cell number density, but in addition, can detect changes in several other physical properties of cells including average shape, average dielectric structure, and size and shape distributions. Thus, differential light scattering can distinguish the difference between a change in light-scattered intensity caused by a change in cell population vs. a change in cell size, structure, and/or distribution.

III APPLICATION OF PRINCIPLE

A. Antibiotic Susceptibility Testing

The precise measurement of the susceptibility or resistance of a culture to an antibiotic is more complicated than the simple detection of viable bacteria because antibiotics may affect both the total bacterial population and/or the bacterial cell itself. These effects include an inhibition of bacterial growth and interference with cell metabolism such that large, small, or even bizarre forms of cells may occur. This is an important distinction in antibiotic testing. An increase in cell mass due to bacterial growth indicates resistance to the antibiotic, whereas if the increase in cell mass is due to greater cell size with no increase in cell numbers, this indicates bacteriostatic effects, implying susceptibility. Thus, an undifferentiated increase in cell mass may give information which may be interpreted incorrectly as indicating growth, i.e., resistance, when in fact, the culture may be producing an increased cell mass due to larger cells — i.e., susceptibility. Turbidimetry and forward single-angle densitometry can measure increases in cell masses, but cannot distinguish whether it was caused by more bacteria or larger cells; whereas differential light-scattering methods make this distinction with the aid of complex, interpretative mathematics. The additional capability of the differential light-scattering technique to detect more subtle changes in the bacterial cell may well produce less equivocal results.

For example, the effects of antibiotics on the structure of the bacterial cell may be even more sensitive and occur more quickly than the changes in the total population with certain types of cultures. This is important where slow-growing cultures such as

Mycobacterium tuberculosis are being tested or with anaerobic bacteria, where growth may be completely inhibited by the incorrect environment. In addition, certain antimicrobials such as the sulfonamides produce only morphological effects during the first few hours of interaction with the cells. In any event, the presence of an antibiotic to which the bacterium is susceptible may bring about changes in bacterial structure greater than in the control even in the absence of growth, which can be detected by changes in differential light-scattering patterns. In these instances, the early detection of alterations in the bacterial cell size and shape and size distribution by differential light scattering may yield more timely results than other techniques or even give definitive results which cannot be obtained otherwise. These principles are purportedly applied in Differential III® to determination of minimum inhibitory concentration (MIC) of an antibiotic by special mathematical treatment of the data. In addition, DLS methods seem capable of producing a rapid measure of *minimum biocidal levels.*

B. Detection and Enumeration

The detection of viable bacteria in a clinical specimen is based on change in biomass or structure during an incubation period. Differential III® can measure such changes at the level of 5×10^5 bacteria/mℓ,[1] although newer data processing techniques[3] indicate detection possibilities in water at 10^4/mℓ. Enumeration can also be performed with this system on certain bacterial populations as changes in logarithmic scattered intensities at several angles are proportional to the logarithm of bacterial population changes.

C. Identification

Bacterial identification may be approached in several ways, including the fact that the physical characteristics of the cell itself may be sufficiently different to identify the bacterium.[4-6] Differential light-scattering techniques offer the unique capability to measure certain characteristics and changes in single bacterial cells which appear to be sufficiently distinctive to identify certain bacteria. It also seems reasonable to expect the immediate applicability of standard biochemical methods with the various sugars, etc., taking the place of antimicrobials. On this latter basis, identification within 3 hr is predicted.[3]

IV. DESCRIPTION OF SYSTEM — DIFFERENTIAL III®

The newest form[3] of Differential III® is intended primarily for clinical laboratory use for antibiotic susceptibility testing and will be described in the most detail. Until 1978, the system produced only qualitative results corresponding to traditional sensitive, intermediate, or resistant designators. The current version is claimed to be able to produce quantitative MIC results. The construction of the instrument is quite simple, consisting of several components, including a rotating sample tray with cuvettes, laser beam, 15-element scattered-light detector array, signal conditioner, minicomputer, and card punch.

Figure 1 is a picture of Differential III® and Figure 2 illustrates the basic scattering geometry of the fixed array light-scattering instrument. Figure 3 is a schematic drawing showing the relationship of the component parts of Differential III® and the sample flow through the system.

To perform an antibiotic susceptibility test, the bacterial suspensions to be tested, together with their controls, are placed in optically clear cuvettes arranged in a circular tray and a laser beam is directed at each cuvette in turn. A detector array positioned in a plane around the cuvette, with each detector at the same fixed radial distance from

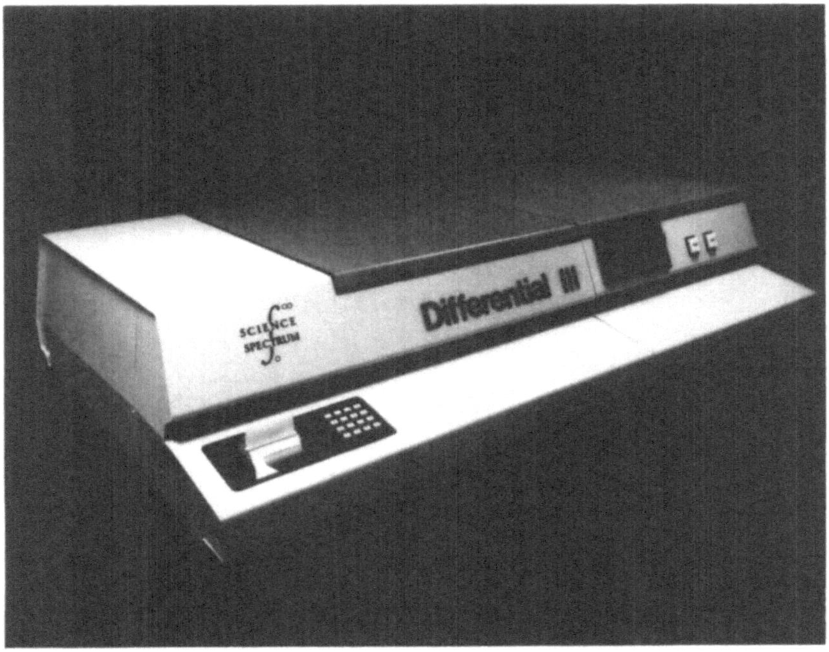

FIGURE 1. Differential III®. (Permission to print granted by Science Spectrum, Inc.)

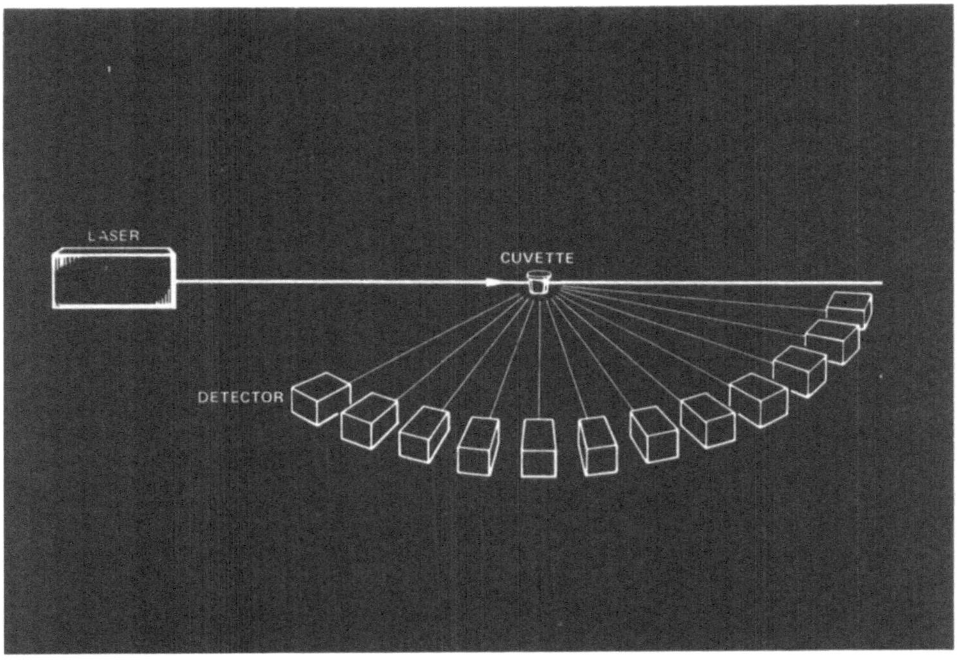

FIGURE 2. A schematic diagram of a differential light scattering measurement. (Permission to print granted by Science Spectrum, Inc.)

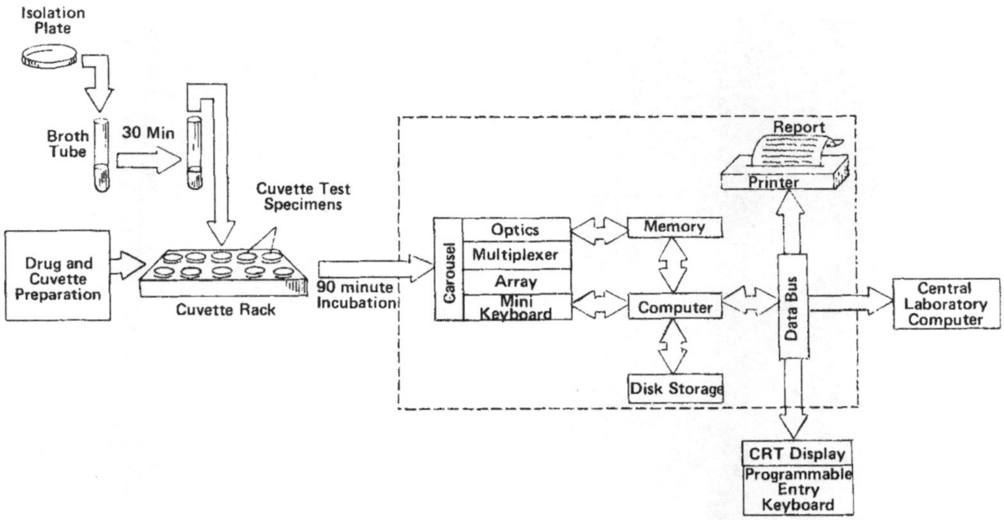

FIGURE 3. Schematic flow chart of sample preparation, Differential III® constituents, and sample progress through the system. (Permission to print granted by Science Spectrum, Inc.)

it, receives the DLS pattern from each cuvette. The 15-element patterns are digitally stored in the computer memory for subsequent processing. Approximately 20 separate sets of array data are processed for each cuvette examined. Each detector is collimated so as to detect only light scattered from a small volume containing a few thousand cells from the center of the cuvette. These data represent an average of the scattering characteristics of the bacteria in the measured volume. The 15-element digitally processed DLS pattern may be subsequently interpolated to yield a more detailed and continuous pattern, as required. Each test pattern is then compared with the control pattern, and a pattern stored earlier of the initial inoculum is compared. The MIC value of each test cuvette is then calculated and recorded on a paper tape. The monochromatic, single mode beam of this laser is linearly polarized, permitting differential light scattering to be studied as a function of polarization. Vertical and horizontal polarized incident light produces quite different patterns, with vertical polarization usually producing the most useful patterns. Vertical polarization is used in Differential III® to produce the DLS patterns.

The Differential III® is equipped for data input and program storage on a small computer disc and is programmable in BASIC for the clinical user. Thus, the MIC values may be modified to yield the simpler sensitive/resistant results or even converted into a customized chemotherapy regimen for treatment.

Differential I® is essentially a manual version of Differential III®, producing only analogue DLS patterns without interpretation. Differential II® is intended for research studies of single particles or cells.[7] Neither Differential I® nor Differential II® is intended for clinical use.

V. METHODS

A. Antibiotic Susceptibility Test (MIC)[3]

Several colonies from an 8 to 18 hr-old plate of the pure culture of the microorganism to be tested are inoculated into broth adjusted to an optical density of 0.1 to 0.15 at 633 nm and incubated at 38°C for 30 min to initiate rapid metabolism (*Pseudo-*

monas may require 60 min). Cuvettes are prepared containing dilute broth and the selected antibiotic impregnated discs yielding specific concentrations, usually in the 1 μg/mℓ range inoculated with 0.3 mℓ of the culture. After capping the cuvettes one cuvette is read and its DLS pattern stored in the Diff III® for future reference. The set is then incubated at 38°C for 90 min, placed into the Differential III®, and read to determine the differential light-scattering pattern from which the MIC values are calculated.

The DLS comparisons and subsequent computation of the MIC values involve the control DLS pattern, the initial inoculum DLS pattern, and the DLS pattern of the treated sample.

These changes in cell characteristics are often related to a known mode of action of the antibiotic, such as with penicillin-sensitive organisms in which cell wall synthesis is inhibited by the drug weakening the wall. Such an example undergoes a shift, causing the cells to swell before eventual lysis in the hypotonic environment. The DLS pattern relates directly to the greater average cell size of the susceptible cells.

B. Identification

Characteristics obtained from single bacterial cell measurements can be used for the identification of the microorganism.[6] This forms the basis of bacterial identification using Differential II®, although this procedure is not yet geared to the routine clinical laboratory. Rapid identification using conventional biochemical tests in combination with DLS measurement is under study by the system developer, although it has not yet been formally introduced.

C. Detection and Enumeration

Reference 6 described details of the DLS approach to these determinations, but at this time no claims are made using Differential III®.

VI. INSTRUMENT FEATURES

Differential III® is used to determine antimicrobial MIC values of microbial specimens, using an automation process of the DLS measurement technique.

These instruments have several important features including (1) fixed 15-element detector array, polarized He-Ne (λ = 633 nm); (2) a laser beam as a light source; (3) a self-contained programmable (in BASIC) computer; and (4) a flexible disc mass storage unit. All of these features aid in the precise measurement and interpretation of the DLS pattern.

VII. INSTRUMENT REQUIREMENTS (DIFFERENTIAL III®)

The instrument requirements are as follows:

1. The instrument requires the specially configured cuvettes designed for Differential III®.
2. An incubator for the cultures is required for the test and is not supplied with the instrument.
3. Antibiotic susceptibility tests require a pure culture and elution discs.

VIII. MARKET STATUS

The Differential III® has been developed and is now in formal clinical trials of its

MIC and synergy determination capability. FDA approval is required for marketing. Until December 1977, Becton-Dickinson and Company was exclusive U.S. licensee. The program has now reverted entirely to Science Spectrum. Once cleared by FDA, the system will be available on a rental-per-MIC basis only at about $0.30 per MIC.

IX. GENERAL COMMENTS

An independent study comparing the results from the earlier qualitative Differential III® and Kirby-Bauer tests was conducted.[2] The correlation of Differential III® and Kirby-Bauer results realized during this independent laboratory evaluation is good. The discrepancies were checked by the tube dilution technique and generally agreed with Differential III® results. No trial data of the MIC determination capability have been released at this time.

X. PRACTICAL CONSIDERATIONS

A. Sensitivity of Different Photometric Procedures

The sensitivity of the different photometric procedures to detect the level of change in the bacterial population to antibiotics varies. In the determination of antibiotic susceptibility, there is more involved than measuring cell mass, since the antibiotics affect the size and structure of the cell, often before a change in population numbers occurs. In instruments intended for antibiotic susceptibility testing, the greater sensitivity of differential light-scattering techniques may well offer faster and more accurate results, especially in testing slow-growing bacteria, anaerobes, and in tests involving antimicrobials such as the penicillins and sulfonamides that influence the size of bacteria sensitive to such agents. This factor should be considered with respect to the intended use of this instrument in the laboratory and the level of sensitivity required to give adequate results for that use.

B. Method

The method recommends picking several like colonies as an inoculum, which introduces the possibility of selecting colonies that appear to be alike, but which in fact are not the same organism. No procedure in the described method will indicate whether this culture is indeed pure. A parallel plating to check for purity may be indicated, or repeated trials using single colonies.

Much of the information in this chapter is summarized briefly in Tables 1 to 3.

TABLE I

Instrument Designation, Principle of Action and Applications of DIFF III®

Instrument trade name	Sponsoring company	Principle of action	Principle of action applied to: (XX = principal application; X = secondary or potential use)				
			Detection	Enumeration	Antibiotic susceptibility	MIC	Identification
Differential III®(also known as Diff III®	Science Spectrum 1216 State St. P.O. Box 3003 Santa Barbara CA 93105	Differential light scattering from bacterial cells showing changes in either cell sizes, structure, size distribution or cell number	X Increase in biomass measured by differential light scattering	X Increase in measured differential light scattering proportional to increase in bacterial cells	XX Change or lack of change in measurement of differential light scattering related to susceptibility with various antibiotics XSerum levels using special strains and previously derived dose-response curves X Minimum biocidal levels may be possible within 3 hr X Drug supply	XX As in previous column, generally with a single concentration of antimicrobial	X Biochemical tests based on rapid DLS response seems promising for 3 hr or less

TABLE 2

Requirements and Features of DIFF III®

	Requirements			Features					
Instrument trade name	Pure culture	Special media	Other	Degree of automation	Test period	No. tests per day per unit	Accuracy compared to conventional tests	Strengths	Weaknesses
Differential III® (Diff III®)	Yes	Perhaps	Special cuvette available from various sources	Semiautomated	30 min + 90 min, read time 10 sec/MIC	1500 to 2500 MIC values per 8-hr day	Good vis a vis Kirby-Bauer. No data yet available re MICs	Differentiates cause of DLS changes. Effects on size, structure, and size distribution readily seen. Programmability of system useful for laboratory integration. Flexibility of test. No fixed number of MIC tests required. Rapid testing even for fastidious organisms. Mass data storage.	Many manual procedures required; pure culture from multicolony inoculum for susceptibility tests and identification

TABLE 3

Market Aspects of DIFF III®

Instrument trade name	Clinical trials	FDA status	Instrument price: Sale $	Instrument price: Lease/rent $	Consumables cost: Per vial $	Consumables cost: Per test $	Commercial potential
Differential III® (also known as Diff III®)	No	On MIC and synergy determination	?	0.40 for 1st MIC; 0.20 each additional set	0.10 times number of drugs	—	Many applications both clinical, industrial, and research to study microparticles in suspension

REFERENCES

1. **Wyatt, P. J.**, Differential light scattering techniques in microbiology, in *Methods in Microbiology*, Vol. 8, Norris, J. R. and Ribbons, D. W., Eds., Academic Press, London, 1973, chap. 6.
2. **Wyatt, P. J.**, Automation of differential light scattering for antibiotic susceptibility testing, in *Automation in Microbiology and Immunology*, Héden, C. G. and Illéni, T., Eds., John Wiley & Sons, New York, 1975, chap. 18.
3. **Wyatt, P. J.**, *Differential light scattering, Rapid Quantitative Antimicrobial Testing and the New Programmable Differential III*, Science Spectrum, Inc., Santa Barbara, California, 1978.
4. **Wyatt, P. J.**, Differential light scattering: a physical method for identifying living bacterial cells, *Appl. Opt.*, 7, 1879, 1968.
5. **Wyatt, P. J.**, Identification of bacteria by differential light scattering, *Nature (London)*, 22, 1257, 1969.
6. **Wyatt, P. J.**, The detection and identification of bacteria by laser light scattering, in *Handbook on Disease Methodology 2*, Amsterdam, D., Ed., Marcel Decker, New York, 1978.
7. **Wyatt, P. J. and Phillips, D. T.**, Structure of single bacteria from light scattering, *J. Theor. Biol.*, 37, 493, 1972.

Chapter 9

AUTOMICROBIC SYSTEM (AMS)®

I. INTRODUCTION

The AutoMicrobic System (AMS)® is a computerized instrumented system with automated processes for the detection, enumeration, and identification/grouping of microorganisms by monitoring changes in optical density through selected media by photometric methods. This instrument system has principally been applied clinically to microbiological analysis of urine specimens. It has been developed, built, and trade-named by McDonnell-Douglas Corporation, St. Louis, Mo. and manufactured by Vitek Systems, Inc. 595 Anglum Drive, Hazelwood, Mo., with disposables distributed by Fisher Scientific Company, AMS Division, 711 Forbes Avenue, Pittsburgh, Pa. U.S. Patents 4,038,151, 3,957,583, and 3,963,355 have been assigned to McDonnell-Douglas Corporation.

II. FUNDAMENTAL PRINCIPLE OF ACTION

Bacteria growing and/or metabolizing in culture media often change the light transmission characteristics of the media. This may be due to an increase in biomass, resulting in turbidity in the broth, or to a change in the color per se, color intensity, or chemical precipitation of the selective media, brought about by the interaction of media constituents with end products of metabolism of the bacteria or a combination of these factors.[1] The metabolic changes are often accentuated by adding indicators to the selective media that are affected by the expected metabolic products. Inhibitory media can also be formulated to selectively prevent the growth of all but one type/group of bacteria, resulting in a growth/no growth situation, which can be monitored for changes in light transmission.

III. APPLICATION OF PRINCIPLE

A. Detection and Enumeration

Using AMS® the detection of bacteria in a specimen is based on monitoring changes in light transmission in the media inoculated with the specimen, and enumeration depends on the rate of change produced by a known dilution of the specimen. The time to detect a positive reaction will vary with each organism, as well as the numbers of bacteria present. The AMS® instrumental system has been used primarily for urine specimens and indicates a positive bacteriuria when 7×10^4 bacteria are present per milliliter of the original sample.

B. Identification

Identification of the predominating bacteria is made by monitoring light transmission changes in highly selective or inhibitory media formulated to allow the growth or appropriate media reaction of only one bacterial type or group.[1,2] Changes in light transmission in specific media indicates the identity of the bacteria present.

The determination of the magnitude of the change in light transmission which indicates a significant result varies with each organism/medium combination. This value is established for each combination and is used as a guideline in programming the computer. It has been shown that competing organisms usually give a lower value than the established threshold for the specific organism/medium combination and are thus disregarded by the computer.

C. Antibiotic Susceptibility Tests

Reduction or inhibition of growth in the test culture as a result of addition of an antibiotic can be detected by monitoring the optical characteristics of each cell for the medium changes which may be caused by growth or metabolism. Little or no change in light transmission after inoculation is interpreted as showing sensitivity of the bacteria to the antibiotic, while transmission similar to the control indicates resistance. A susceptibility test for *Escherchia coli* is now available. Therefore, the AMS® can perform all parts of microbiological analysis by monitoring photometric changes in light transmission brought about by test conditions.

IV. INSTRUMENT DESCRIPTION AND OPERATION[3]

The AMS® is a system using automatic processes for the detection, enumeration, identification/grouping, and eventually antibiotic susceptibility testing of bacteria directly from clinical specimens using selective media. It is depicted in Figures 1 and 2. At present, only urine specimens are being tested, and the following discussion is based on urine analysis. Results are measured by transmission of light through cuvettes containing the sample. The system consists of Urine Identification Cards, a sample injector, a diluent dispenser, filling module, reader/incubator module, computer control module, and data terminal.

A. Urine Identification Card Micro Plate (Figure 3)

The Urine Identification Card Micro Plate is a clear plastic disposable micro plate, both faces of which are covered with air-permeable tape forming 20 3-mm diameter conical wells. These wells are of two types, nine with freeze-dried selective media specific for common urinary tract pathogens termed identification wells and five enumeration wells containing nonselective media. A positive control well contains nonselective media to cover urinary pathogens other than the nine specific pathogens. The identification wells contain selective media for

Pseudomonas aeruginosa
Citrobacter freundii
Serratia sp.
Escherichia coli
Klebsiella/Enterobacter
Yeasts
Enterococcus Group D
Staphylococcus aureus

The reactions shown by these media are described in detail in a paper by Aldridge et al.[1] The wells are connected to the sample inlet by distribution channels molded in the plastic.

B. Sample Injector (Figure 3)

The sample injector is a plastic two-chambered device, one short and one long chamber, which is used for making two different dilutions of the urine specimen. Each chamber has an injecting needle capable of piercing the inlet septa of the Urine Identification Card and under vacuum of transferring the less diluted urine to the identification wells and the positive control and the more diluted urine to the enumerating

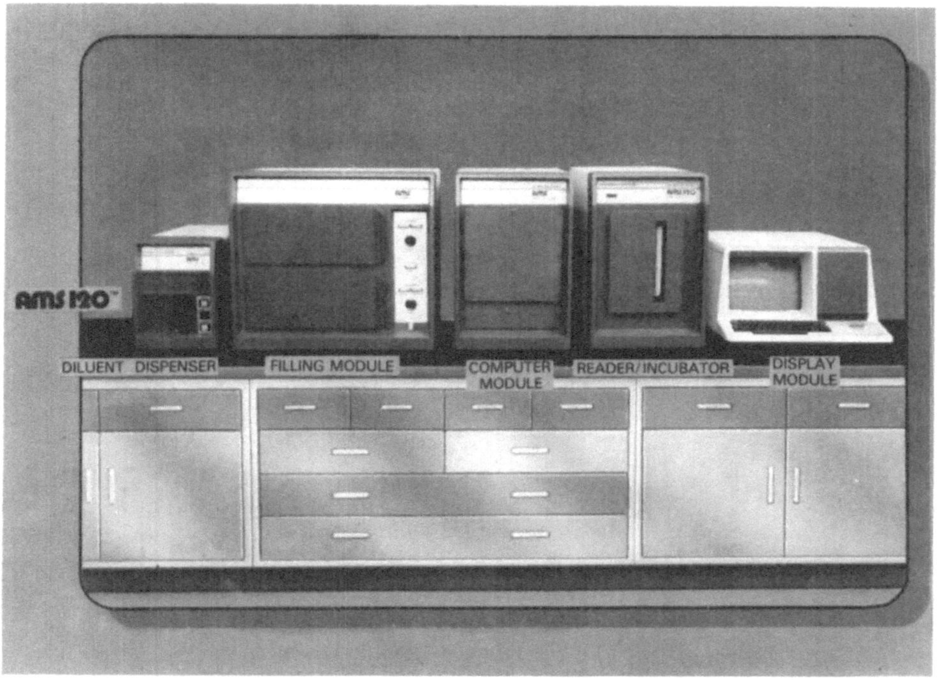

FIGURE 1. Auto Microbic-120 Instrumented System. (Permission to print granted by Vitek Systems, Inc.)

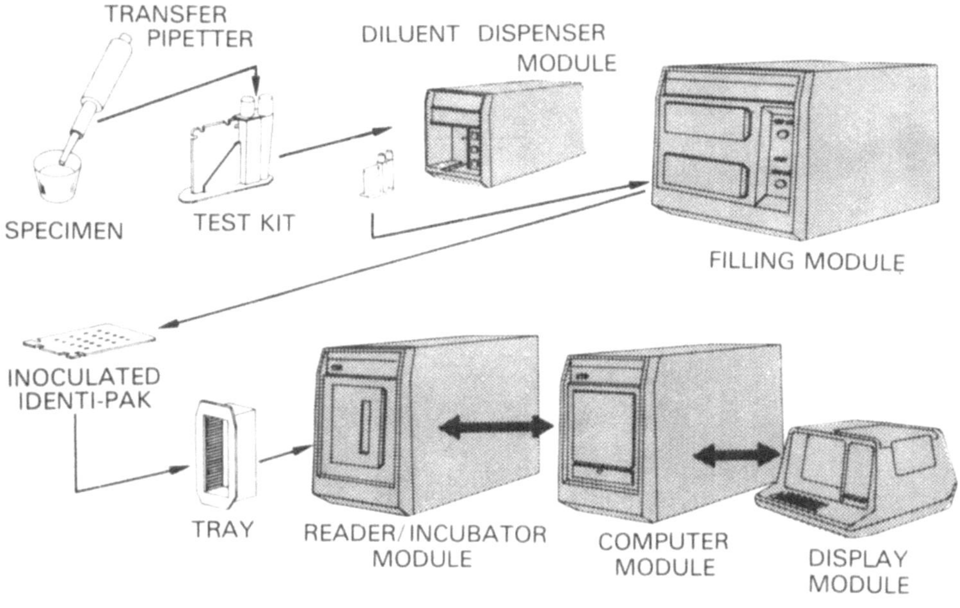

FIGURE 2. Flow diagram of Auto Microbic-120 instrumented System. (Permission to print granted by Vitek Systems, Inc.)

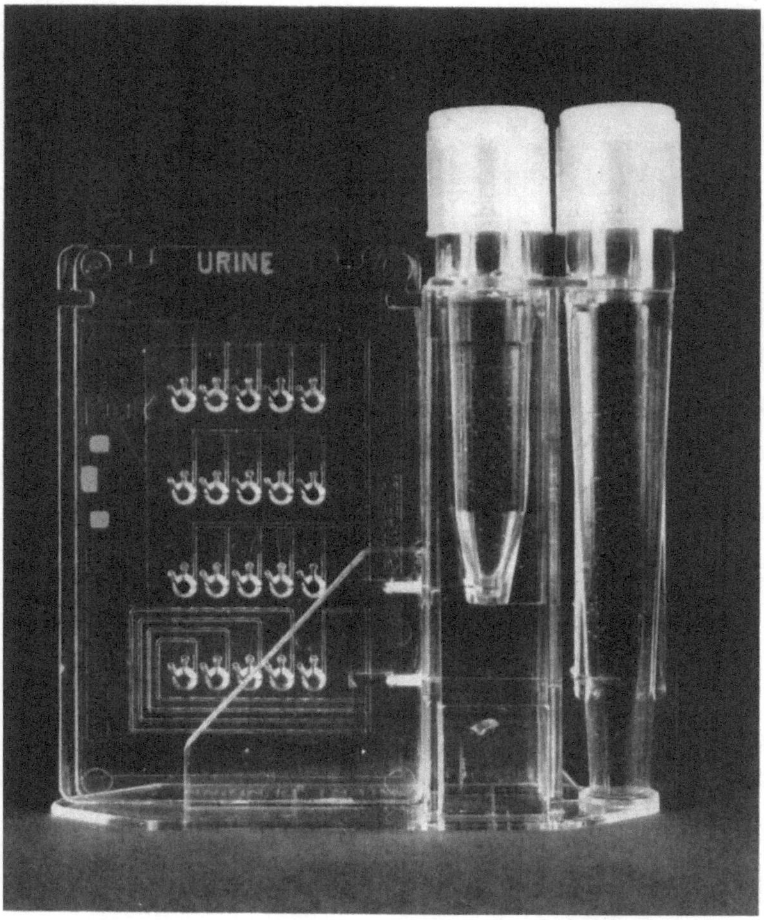

FIGURE 3. Urine Identification Card and Sample Injector for Auto Microbic Systems. (Permission to print granted by Vitek Systems, Inc.)

wells. Proper alignment of the injector and the card during coupling is accomplished by a parallel-walled slideway, and filter caps on the two chambers prevent cross contamination.

C. Diluent Dispenser

This device makes the proper dilutions of urine by discharging the saline into the two chambers under pressure to ensure proper mixing of diluent and urine.

D. Filling Module

The filling module receives the racks of ten coupled Urine Identification Cards and the sample injector with the diluted urine and by vacuum transfers the urine from the injector to the card.

E. Reader/Incubator Module

Incubation at 35°C and reading at hourly intervals is carried out in the reader/incubator on the wells in vertical 30-position trays. Each trayload is scanned for 6 min each hour by an electro-optical scanning device, which reads the wells in a pattern that eliminates certain incidental anomalies. Light-emitting diodes sense the growth indices and the sample identification number when the plate is moved into the reader.

F. Computer/Control Module

The computer with a 4-K core memory for data buffering and storage is coupled with a programmable memory with 5-K capacity for program instructions and an electronic interface in the computer/control module. There is a fail-safe system to pinpoint errors.

G. Data Terminal

The data terminal consists of a display, keyboard, and electrostatic printer which automatically records the final data on samples which have completed the 13-hr test and also provides data on command.

V. METHODS

A. Identification/Grouping and Enumeration

To perform identification/grouping and bacterial enumeration on a urine specimen using AMS® the following procedure is recommended.[3]

1. Mark specimen ID on Urine Identification Card.
2. Couple card with sample injector so that the injector needles penetrate the inlet septa of card.
3. Remove filter caps, pipette 0.2 mℓ urine into the short chamber and 0.005 mℓ into the long chamber of the sample injector and add 1.9 mℓ 0.5% saline to the short chamber and 5 mℓ diluent to the long chamber, using pressure to ensure mixing. This gives a 1:10 dilution in the short and 1:1000 in the long chamber. Replace the caps and place rack of coupled card injectors into the filling module.
4. Produce vacuum in filling module for 5 min, remove rackload of test samples from module, discard sample injector, and place Identification Cards in a tray into the carrousel of the reader/incubator, not necessarily in sequence. Trays may be placed in the module at any time except during the 6-min reading cycles.
5. The incubation is continuous and readings will be carried out automatically at hourly intervals for 13 hr.
6. The final report is automatically printed out at 13 hr and the results can then be reported as normal (negative for bacteriuria) or positive, indicating 70,000 organisms or more per milliliter, which requires medical follow-up.
7. The grouping/identification will be reported based on the designated well in which growth was detected.

B. Susceptibility Testing

Specimens that show 70,000 organisms/mℓ or more on the Urine Identification Card are candidates for antibiotic susceptibility tests, which may be performed using the following procedure.[3]

1. Note the identification of the organism in the positive well and select a Susceptibility Card appropriate for that organism.
2. Mark the card with the specimen ID. Inoculate this Susceptibility Card with the contents of positive well on the Urine Identification Card by penetration of the taped wall of the well with a needle, withdrawing the contents, diluting it 1000 times, using a single-chambered sample injector, and transfer the inoculum as before into the Susceptibility Card.

3. Insert the Susceptibility Card micro plate into the reader/incubator module, incubate for 12 hr and read.
4. The automatically printed report will show R for resistant or S for susceptible. No procedure for MIC determinations is offered in the product literature. The susceptibility test for *E. coli* is now available.

VI. INSTRUMENT FEATURES

Instrument features are as follows:

1. The instrument processes the specimen automatically after pipetting urine and placing Identification Cards into the filling module and transferring cards to the reader/incubator in tray.
2. The system accepts the diluted original urine specimens and can perform a complete microbiological analysis without requiring pure cultures. The positive cultures from the detection identification card may be processed for antibiotic susceptibility by transferring the diluted contents of the positive well of the Identification Card to the single-chamber sample injector for inoculation into the Susceptibility Card. To prevent the growth of a second organism which may be present in the urine, the same selective media contained in the positive well is in the Susceptibility Card wells, ensuring that only the one identified organism will grow in that susceptibility test.
3. The system calibrates itself automatically.
4. The instrument is available in two sizes: Model 120 AutoMicrobic System, which accepts 120 micro plates, and Model 240 AutoMicrobic System,which accepts 240 micro plates.

VII. INSTRUMENT REQUIREMENTS

Instrument requirements are as follows:

1. The Urine Identification Cards and the Susceptibility Cards with the proprietary selective media, together with the sample injector must be used in this instrument system and must be stored at 0 to 5°C.
2. A filling module must be used for inoculation.

VIII. MARKET STATUS

The AutoMicrobic System is being sold presently on its urine sample testing capability. The susceptibility tests for *E. coli* and *Proteus* have passed FDA certification and are now on the market. Biochemical identification of Enterobacteriaceae is now available. Now awaiting approval to market are: Yeast Identification Card, General Susceptibility Card (Gram-negative), MIC Susceptibility Panels I and II (Gram-negative). Other susceptibility tests are still awaiting FDA approval. Early in 1977, a demonstration/evaluation program was undertaken, which has resulted in over 20 in-laboratory testing programs to date. Under this program, microbiology laboratories are given an opportunity to see the AMS® perform under their own laboratory conditions. Systems have now been sold, rented, and leased throughout the U.S. and Canada.

IX. GENERAL COMMENTS

The AutoMicrobic System was developed by the McDonnell-Douglas Corporation. The project originated with the company's proposal to National Aeronautics and Space Administration (NASA) to create an automated microbiology system to meet space exploration requirements. This led to the development of a totally separate system directed toward clinical use — the AutoMicrobic System.

In early tests with urine specimens, the AMS® showed favorable results with most organisms.[1] The new methodology is being subjected to exhaustive study for validation and correlation with established techniques. This on-going collaborative program has proceeded in 1975, 1976, and 1977 with the cooperation of six of the foremost clinical microbiology laboratories in the U.S. serving as independent investigators. [4,5] The results of the program have been very favorable. A developmental study on the use of AMS® for antibiotic susceptibility testing on urine cultures of five common urinary pathogens showed good correlation with Kirby-Bauer tests.[6,7]

In a recent report by Marso[8] on 600 clinical urine specimens and by Madden[9] on 6000 urines tested to compare traditional methods and AMS® procedures, there was good correlation both with respect to detection of bacteriuria and identification of the bacteria, as well as antimicrobial susceptibility of *E. coli.*

X. PRACTICAL CONSIDERATIONS

AMS®, when fully developed and approved, offers the potential to do a complete microbiological analysis. One of the principal purposes of automated systems for microbiological assays is to speed up the availability of the analytical results to the doctor. In considering the addition of an instrumented system to any laboratory, the question of meeting this goal must be considered. At present, the AMS® takes up to 13 hr to detect a bacteriuria and group-identify its predominating bacteria. To determine susceptibility takes additional time (about 12 hr), and since this test is not yet available, the only recourse at present is to transfer the positive well contents and perform a routine Kirby-Bauer susceptibility test. This raises the question of how much sooner a patient will receive treatment using AMS®, as it is now approved for clinical laboratory use. Marso[8] reported that the AMS® did yield significantly faster reportable results in his hospital.

One of the questions still to be answered is the accuracy of certain of the antibiotic susceptibility tests as a treatment guide. The susceptibility test as set up with AMS® makes use of essentially the same basal media for the antibiotic tests as was present in the positive well from which the inoculum is taken. This means that in certain tests the possibly inhibitory media used to suppress all but the designated type of bacteria is also used in the susceptibility test. Since it appears that the selectivity of some of these media is based in part on the inclusion of other bacterial inhibitory substances, it means the susceptibility of the bacterium under test is being evaluated for the test antibiotic plus other inhibitory substances in the test media which will not be used in conjunction with it in the treatment of patients. The effect of these additional inhibitors on the susceptibility results must be carefully evaluated for accuracy when these results are used as a treatment guide.

Much of the information in this chapter is summarized briefly in Tables 1 to 3.

TABLE 1

Instrument Designation, Principle of Action, and Applications of AMS®

Instrument trade name	Sponsoring company	Principle of action	Principle of action applied to: (XX = principal application; X = secondary or potential use) Detection	Enumeration	Antibiotic susceptibility	MIC	Identification
AutoMicrobic system (AMS)®	McDonnell-Douglas Corp., E. St. Louis, Mo. Manufactured and distributed by Vitek Systems, Inc., 595 Anglum Dr., Hazelwood, Mo. 63042 Distribution of disposables, Fisher Scientific Co., ANS Division, 711 Forbes Ave., Pittsburg, Pa.15219	Changes in light transmission through culture in selective media	XX Change in turbidity or color per se, color intensity or chemical precipitation of selective media or combination of above	XX As in previous column, rate of change proportional to increase in bacterial population	XX X *E. coli* Increase or lack of increase of transmitted light measurements related to biomass using various antibiotics	X As in previous column, but using various concentrations of one antibiotic	XX Distinctive pattern of changes in light transmission as noted in detection column

TABLE 2

Requirements and Features of AMS®

	Requirements			Features					
Instrument trade name	Pure culture	Special media	Other	Degree of automation	Test period	No. tests per day per unit	Accuracy compared to conventional tests	Strengths	Weaknesses
AMS® 120	No	Yes	AMS® line of expendables including cuvette, sample injector, filling module	Automated after sample preparation and inoculation	13 hr detection/identification, and 12 additional hours for susceptibility tests	120	Good	Does not need pure culture; potential for complete microbiological analyses when fully developed and approved	Length of detection time; use inhibitors in antibiotic susceptibility test media
AMS® 240						240			

TABLE 3

Market Aspects of AMS®

Instrument trade name	Clinical trials	FDA status	Instrument price		Consumables cost		Commercial potential
			Sale	Lease/rent	Per vial	Per test	
AMS® 120	Yes, 6 laboratories	Approved bacteriuria tests and susceptibility testing for *E. coli* and *Proteus*. Approved biochemical ID of Enterobacteriaceae; Pending Yeast ID, General Susceptibility (Gram Neg), MIC Susceptibility Panels I and II (Gram Neg).	$49,500	$1500	$3.00	$3.00 (detection and identification) $3.00 (antibiotic susceptibility)	Method can be extended to clinical specimens other than urine.
AMS® 240			$69,500	$2100	$3.00	$3.00	

REFERENCES

1. **Aldridge, C., Jones, P. W., Gibson, S., Lanham, J., Meyer, M., Vannest, R., and Charles, R.,** Automated microbiological detection/identification system, *J. Clin. Microbiol.*, 6, 406, 1977.
2. **Sonnenwirth, A. C.,** Preprototype of an automated microbial detection and identification system: a developmental investigation, *J. Clin. Microbiol.*, 6, 400, 1977.
3. AutoMicrobic System Bulletin Fisher Scientific Co., Pittsburgh, Pa. #382/2-720-120.
4. **Smith, P. B., Balows, A., Washington, J., Isenberg, H., Gavin, T., Sonnenwirth, A., Taylor, W.,** A collaborative evaluation of the AutoMicrobic system, presented at *77th Annu. Meet. Am. Soc. Microbiol.*, New Orleans, 1977, 52.
5. **Isenberg, H. D., Telenson, M., Washington, J., II, Sonnenwirth, A. C., and Taylor, W.,** Clinical evaluation of the AutoMicrobic System Urine IdentiPak, presented at *78th Annu. Meet. Am. Soc. Microbiol.*, Las Vegas, 1978, 302.
6. **Fadler, N., Jones, P. W., and Gibson, S.,** Developmental study of automated susceptibility testing using the AutoMicrobic System, presented at *77th Annu. Meet. Am. Soc. Microbiol.*, New Orleans, 1977, 40.
7. **Isenberg, H. D., Scherer, J., Bajan, S., and Fox, C.,** Initial clinical laboratory evaluation of AMS susceptibility testing, presented at *17th Interscience Conf. Antimicrobial Agents and Chemotherapy*, New York, 1977, 384.
8. **Marso, E. L., Cue, J., Oshima, C., Wilkenson, V., and Martin, W. J.,** Comparison of the AutoMicrobic System with conventional laboratory methods for quantitation, identification and susceptibility testing of urinary tract isolates, presented at *78th Annu. Meet. Am. Soc. Microbiol.*, Las Vegas, 1978, 301.
9. **Madden, J., Higbee, J., and Johnson, R.,** Clinical evaluation of the McDonnell-Douglas AutoMicrobic system, presented at *78th Annu. Meet. Am. Soc. Microbiol.*, Las Vegas, 1978, 302.

Chapter 10

MS-2® INSTRUMENTED SYSTEM

I. MS-2®

The MS-2® (trade name — Abbott Laboratories Diagnostics, Division, 4757 Irving Boulevard, Dallas, Texas), using automatic processes, performs antibiotic susceptibility tests turbidimetrically. Two U.S. patents, 3,837,745 and 3,983,006, are assigned by J. L. Acker and Peter Meserol to Akro-Medic Engineering, Inc., Denville, New Jersey.

II. PRINCIPLE OF ACTION

MS-2® uses densitometric measurements to determine growth or inhibition of growth in bacterial cultures in antibiotic susceptibility tests.

A turbidimetric assay is a straightforward method that measures the light that passes through a suspension of particles by placing the light source on one side and the photodetector on the other side so that the light passes directly through the suspension to be tested. The method is sensitive enough to detect a change in bacterial cell mass, so it can be used in microbiological analyses which depend on changes in biomass. Less light is transmitted as the biomass, i.e., the number or size of the bacterial cells, increases, but this technique does not differentiate between an increase in biomass caused by more particles vs. larger particle size.

III. APPLICATION OF PRINCIPLE

A. Antibiotic Susceptibility Testing

Antibiotic resistance or susceptibility may be expressed by the growth or lack of growth of a bacterial culture exposed to that antibiotic, resulting in an increase (resistance) or lack of increase (susceptibility) in turbidity. MS-2® uses a sensitive photometer that measures turbidity to determine susceptibility or resistance. In addition, by calculations from the susceptibility data, MICs can be determined.

B. Detection and Enumeration

Applying the same principle, growth of bacteria from a specimen can be detected by MS-2® by an increase in turbidity and enumeration can be performed as an increase in turbidity is directly related to the bacterial population. The minimal concentration of bacteria detected is $10^5/m\ell$.

C. Identification

MS-2® has the potential to identify bacteria by using lyophilized biochemicals in differential media and detecting end points.

IV. INSTRUMENT DESIGN AND OPERATION[1,2]

The MS-2® is a modular instrumented system performing automated processes (Figure 1) designed to determine antibiotic susceptibilities and minimum inhibitory concentrations (MICs) by kinetic analysis of microbial growth using the principle of turbidimetric assay. The system is composed of four components — the cuvette cartridge, the disc loader/sealer, the analysis module, and the computer module.

FIGURE 1. MS-2® Instrument System. (Permission to print granted by Abbott Diagnostics Division.)

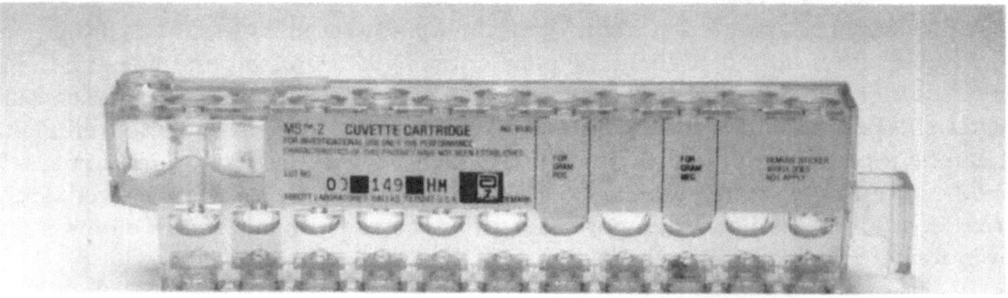

FIGURE 2. Cuvette for MS-2® Instrument System. (Permission to print granted by Abbott Diagnostics Division.)

A. Cuvette Cartridge (Figure 2)

The cuvette cartridge which serves as the test vehicle is a disposable, plastic 11-chambered assembly. The cuvette cartridge has a unique construction which allows communication of the 11 optical chambers with 2 common upper chambers through special hydrophobic membranes. The culture is introduced in the upper chamber, incubated, and monitored for growth. When log phase of growth is attained, the broth culture is transferred by vacuum through the membrane floor into the 11 chambers, 10 of which contain antibiotics, and the 11th serves as a control.

B. Disc Loader/Sealer

The disc loader/sealer is a mechanized device for dispensing the antibiotic discs into the bottom chambers and using high pressure to seal the chambers prior to inoculation.

C. Analysis Module

The analysis module represents a complex optical analyzer that receives, incubates at 35°C, linearly agitates, and monitors at 5 min-intervals eight inoculated cuvette cartridges. Each cuvette has its own individual optical system consisting of light-emitting diodes and matched detectors.

D. Computer Module

The computer module controls the functions of the analytical modules and the optical systems. Optical density, generation time, growth rate constants, and comparative inhibition of microbial growth are calculated from the data from as many as seven analysis modules by a single compact computer module using solid state microprocessors. Results are recorded both by a high-speed thermal printer and on magnetic tape for data storage.

To operate the MS-2,® the antibiotic-filled, sealed, and inoculated cuvette cartridge is loaded into the analysis module, after which all subsequent operations are automatically controlled by computer.

V. METHODS[1]

A. Antibiotic Susceptibility Tests

At present the only claims for MS-2® are for antibiotic susceptibility and MIC tests.

1. The cuvette cartridges are loaded with ten appropriate antibiotics selected for Gram-negative, Gram-positive, or urinary tract bacteria, sealed, and the upper chamber inoculated with several representative colonies from a Petri plate, which are vortexed and suspended in modified Oxoid Sensitest broth.®*
2. The loading port is closed, and the cartridge is inserted into an assigned position in the analysis module.
3. All remaining operations are automatic. The upper chamber is scanned until the culture reaches logarithmic growth and is transferred to the lower antibiotic-containing and control cuvettes. Each cuvette is scanned every 5 min, and the data is stored for mathematical comparison with later readings over a number (four to five) of generations, giving a true comparative kinetic analysis. The computer calculates and prints out the final analytical data for each isolate.

   The types of data produced by the MS-2® can best be visualized by making graphic plots of the growth curves from which the computer derives its analysis.
4. One of the advantages of kinetic analysis is that the antibiotic may produce different effects as growth of the culture progresses. Some antibiotics inhibit growth or cause lysis only during certain stages of growth. A single reading taken before inhibition of growth may lead to an erroneous resistant interpretation.
5. MIC — the MS-2® routinely calculates through the use of appropriate algorithms quantitative clinically relevant minimal inhibitory concentration (MIC) within attainable dosage ranges from tests performed as described above, where the antibiotic produces partial inhibition of growth at the tested concentration. From these data, the computer can calculate the concentration of antibiotic required to produce complete inhibition (MIC), usually in 2 to 6 hr, depending on the growth rate of the test organism.[2,3]

B. Detection, Enumeration, and Identification

No published methods are available for the use of MS-2® to detect, enumerate, or identify bacteria.

VI. INSTRUMENT FEATURES

1. After inoculation of the colonies into broth, the instrument is completely automatic.

* Manufactured for Abbott Laboratories by Oxoid Ltd., Wade Road, Basingstoke Hampshire RB240PQ, England.

2. MS-2® automatically monitors the growth of the prepared test culture until the proper predetermined turbidity is reached, at which time the test culture is added automatically to the antibiotic-containing chambers and their control. This ensures the proper concentration of rapidly growing organisms for the test.
3. MS-2® routinely provides MIC for more meaningful guidance in treatment.

VII. INSTRUMENT REQUIREMENTS

1. The unique MS-2® cuvette cartridge is required for tests in the MS-2® analysis module.
2. The disc loader/sealer is specially configured to load the cuvette cartridge.
3. MS-2® Culture Medium is especially formulated for the tests with MS-2®.
4. The procedure requires a pure culture for test.

VIII. MARKET STATUS

The MS-2® test system has received approval for sale from U.S. FDA for antimicrobial susceptibility testing, bacterial identification and urine screening for bacteriuria.

IX. GENERAL COMMENTS

Stability and reproducibility of results are important aspects of any instrumented system, and experiments were designed to evaluate these factors.[1] The results showed good uniformity of inoculum input and response of the MS-2® system plus a high degree of reproducibility. Such findings are encouraging, since one objective of automation in microbiology is to minimize variations due to subjective differences in evaluation.

A preliminary study on antibiotic susceptibility tests on 200 clinical isolates showed good correlation of results from MS-2®, both Kirby-Bauer and broth dilution methods, as well as good reproducibility and accuracy.[4,5]

A collaborative study of the MS-2® system is underway in four major laboratories in the U.S., with the findings to be jointly published at the conclusion of the study. Results of clinical evaluations to date are promising.[6]

Potentially the MS-2® can perform organism identification utilizing lyophilized biochemicals and detecting end points.[7]

X. PRACTICAL CONSIDERATIONS

There is increasing recognition that MICs offer more information as a treatment guide than qualitative susceptibility values, and the MS-2® offers MIC values routinely. If MICs are in demand in a laboratory, this factor should be considered in acquiring an instrument.

In traditional systems, the manual standardization of the inoculum prior to starting the test is time consuming and sometimes inaccurate. After a simple preparatory step, the automatic monitoring of the test culture for the proper turbidity prior to the automatic addition of the test culture to the antibiotics as performed by MS-2® may be a deciding feature in selecting this instrument because it ensures an accurate concentration of bacteria in the log phase of growth without manual intervention.

The method recommends picking several like colonies as an inoculum, which introduces the possibility of selecting colonies that appear to be alike, but which in fact are not the same organism. As no procedure in the described method will indicate whether this culture is indeed pure, a parallel plating to check for purity may be indicated.

One of the contributions that instrumented microbiological analysis can make is to extend the knowledge of microbial reactions. Thornsberry et al.[8] describe an important study on the MS-2® to yield graphs showing optical density vs. time, comparing the growth curves of the control and antimicrobial-containing cultures. These graphs give a visualization of the kinetics of the drug/organism interaction, as well as delayed drug activity, bacteriolytic action, and emergence of mutants. This helps to explain that certain early susceptible reactions to an antibiotic are due to bacteriolysis, but on longer incubation growth may re-emerge. Since standard methods reflect only the latter condition, the early susceptible results would not agree with the traditional method. (Editor's note: this raises an interesting question. Which result is closer to the reaction taking place in the body, i.e., which is the better guide for therapy? Is it the test showing the action of the antibiotic on the log phase culture or on the stationary phase of growth?) (See the discussion in Chapter 3.)

Much of the information in this chapter is summarized briefly in Tables 1 to 3.

REFERENCES

1. **Spencer, H. J., Stockert, J., Welaj, P., Wilborn, R., and Price, B.,** Automated antimicrobic susceptibility testing with the MS-2® system, in *2nd Int. Symp. Rapid Methods and Automation in Microbiol.*, Johnston, H. H. and Newsom, S. W. B., Eds., Learned Information Ltd., Cambridge, England, 1976, 272.
2. **Stockert, J. E., Wilborn, R. C., and Fekete, N. M. G.,** Rapid automated susceptibility testing. Test instrument description (MS-2), presented at *77th Annu. Meet. Am. Soc. Microbiol., New Orleans,* 1977, 56.
3. **Valu, J., Meserol, P., and Acker, J.,** Rapid sequential bacteriological analyzer (RSBA) predictions of MIC of cephalothin for *Escherichia coli* clinical isolates, presented at *16th Interscience Conf. Antimicrobial Agents Chemother.*, Chicago, 1975, 316.
4. **Spencer, H. J., Welaj, P. M., and Swensen, L. M.,** Rapid automated antimicrobial susceptibility testing. Comparison of results from automated instrument system (MS-2) with standard methods, presented at *77th Annu. Meet. Am. Soc. Microbiol.*, New Orleans, 1977, 57.
5. **Spencer, H. J., Welaj, P., Vannest, R., Hargrove, V., and Williams L.,** Accuracy and reproducibility of the MS-2 automated susceptibility testing system, presented at *78th Annu. Meet. Am. Soc. Microbiol.*, Las Vegas, 1978, 311.
6. **McCarthy, L. R., Sherris, J. C., and Anhalt, J. P.,** Collaborative evaluation of the MS-2 antibiotic susceptibility testing instrument, presented at *78th Annu. Meet. Am. Soc. Microbiol.*, Las Vegas, 1978, 311.
7. **Mayfield, D. C.,** personal communication.
8. **Thornsberry, C., McCarthy, L. R., and Sherris, J. C.,** Antimicrobial susceptibility patterns for bacteria determined by visual analysis of computer generated growth curves, presented at *78th Annu. Meet. Am. Soc. Microbiol.*, Las Vegas, 1978, 297.

TABLE 1

Instrument Designation, Principle of Action, and Applications of MS-2®

Instrument trade name	Sponsoring company	Principle of action	Principle of action applied to: (XX = principal application; X = secondary or potential use)				
			Detection	Enumeration	Antibiotic susceptibility	MIC	Identification
MS-2®	Abbott Diagnostics Division, Dallas, Texas	Turbidimetric growth rate analysis	X (Turbidimetric) light-emitting diodes with matched detectors at 670 nm wavelength	X Increases in turbidity directly related to bacterial population	XX Utilizes comparative growth rate analysis to determine resistance or susceptibility	XX Uses linear regression analysis of growth rate constants to determine MIC values	X Potential application utilizing lyphilized biochemicals in an end-point detection

TABLE 2

Requirements and features of MS-2®

	Requirements			Features					
Instrument trade name	**Pure culture**	**Special media**	Other	Degree of automation	Test period	No. tests per day per unit	Accuracy compared to conventional tests	Strengths	Weaknesses
MS-2®	**Requires pure culture isolate**	Requires MS-2® Modified Oxoid Sensitest® Media	Special cuvette; special discs and dispenser	Completely automated after sample introduction into instrument; results automatically printed out upon analysis.	Time frame of 3—5 hr per test	Range of 16—24 tests per 8-hr day	Not presently available from collaborative evaluation	Automatic result print-out of R, S, I values plus clinically relevant MIC values can monitor inoculum and inoculate automatically	Uses multicolony inoculum for susceptibility and identification tests

TABLE 3

Market Aspects of MS-2®

Instrument trade name	Clinical trials	FDA status	Instrument price: Sale $	Instrument price: Lease/rent $	Consumables cost: Per vial $	Consumables cost: Per test $	Commercial potential
MS-2®	Yes; 4 on-going* tests	Approved for antimicrobial susceptibility, testing, bacterial ID, and urine screening for bacteriuria	34,000 1 control module 2 analysis modules	Options available	3.50	3.50	Antimicrobial susceptibility testing Urine screening Serum bioassay Bacterial identification

* CLINICAL INVESTIGATORS

1. University of Washington. With John Sherris, M. D. and Fritz Schoenknect, M.D.
2. Center for Disease Control, Clyde Thornsberry, Ph.D.
3. Mayo Clinic, John Washington, M.D. and John Anhalt, M. D.
4. University of North Carolina, Laurence R. McCarthy, Ph.D.

Chapter 11

BACTOMETER® MICROBIAL MONITORING SYSTEM

I. INTRODUCTION

Bactometer® Microbial Monitoring System is the registered trademark of a series of instrumented systems using automated processes for microbiological analysis by the measurement of electrical impedance. The instruments are made by Bactomatic, Incorporated, 1049 Elwell Court, Palo Alto, Calif.

Presently these instruments are principally used for the detection of bacteremia and bacteriuria. They also have the potential for performing antibiotic susceptibility and identification tests. U.S. Patent 3,743,581 was issued to Paxton Cady and William J. Welch and assigned to Bactomatic, Inc. Other patents include U.S. Patents 3,984,766, 3,890,201, and 4,067,951.

II. FUNDAMENTAL PRINCIPLE OF ACTION

Impedance is the resistance to the flow of an alternating electrical current and is analogous to the resistance in a DC circuit. Cultures of microorganisms bring about changes in the chemical composition of the medium in which they grow through the enzymatic activity associated with growth and metabolism. These chemical changes alter the impedance of the medium and provide an indirect measure of microbial growth and metabolism.

Impedance is a complex entity consisting of both conductive and reactive (capacitive) components. Both components are seen to change with microbial growth and contribute to the measured change in impedance.[1,2] For most growing microorganisms, impedance is seen to decrease with time. Electrical measurements have been used to monitor microbial growth since the end of the last century. However, recent interest in this approach for clinical applications stems from the presentation of papers independently in 1973.[3-5]

III. APPLICATION OF THE PRINCIPLE

Impedance monitoring may serve as the basis for a number of microbial analyses in which microbial growth and metabolism play a role. These include detection and estimation of the initial concentration of viable microorganisms, identification, and determination of antibiotic susceptibilities.[1]

To minimize nonmicrobial effects upon impedance measurements from physical and chemical factors, the Bactometer® Microbial Monitoring System series of instruments make measurements using both a sample chamber and an uninoculated reference chamber. The same medium is used in both chambers. In the Bactometer® 8 and 32, each sample chamber is paired with a reference chamber, and the instruments actually measure changes in the impedance ratio. In this way, nonspecific impedance changes occurring in both chambers tend to cancel out. In the Bactometer® 120 Microbial Monitoring System, a single reference chamber is used for each 15-sample chamber, and the instrument's microprocessor compensates for much of the nonmicrobial impedance change.

The major physical factor affecting impedance measurements is temperature, a change of 1°C causing about a 2% change in the impedance. Other factors include evaporation, evolution or absorption of gases, settling of particles from suspension, and the lysis of red blood cells.[6]

A single-sample chamber in any Bactometer® Microbial Monitoring System is measured electrically frequently enough, (e.g., every 96 sec in the Bactometer® 32 system), relative to the rate of microbial growth, to provide a virtually continuous record of the impedance changes caused by microbial growth and activity. Initially, there is a brief period of impedance change corresponding to the establishment of thermal equilibrium as the sampling chambers reach incubator temperature. This is followed by a flat baseline which lasts until the microorganisms replicate to a concentration sufficient enough to cause a detectable impedance change. The time required to reach this point is the detection time, and the concentration of microorganisms at detection is known as the threshold concentration.

With further monitoring, the changes in impedance accelerate, frequently doubling at the same rate as the growth rate (as determined by plate counting), so that the impedance doubling time approximates the generation time. Later, the curve of impedance change with time assumes an S-shape similar to the familiar sigmoid growth curve. In the region corresponding to the stationary phase in a growth curve, the impedance curve often continues to change, even though the population of organisms has ceased to increase; this presumably reflects continued metabolic activity.[2]

The threshold concentration depends in part upon the organism, the medium, the type of electrode, and the definition of detection used. When detection is defined as 0.8% accelerating impedance change from baseline (a practical definition), the threshold for most organisms in most media is between 10^6 and 10^7 organisms per milliliter. When detection is defined as the earliest perceivable change in impedance (which is significantly different from changes due to noise and drift), the thresholds are lower, lying between 10^5 and 10^6 organisms per milliliter.

The detection time is a function of the initial concentration of organisms present in the test sample, the generation time for that population of organisms in the environment being used, and the threshold for that organism/medium/electrode combination. Generally, for cultures with the same growth rates, the lower the initial concentration, the later the detection time. If the detection times are then plotted against the logarithm of the initial concentration, the points should be along a straight line. The detection time can then be used as a rough index of the initial concentration, assuming a fixed initial lag period.

The following microbial analyses may be applied clinically to body fluids such as blood, urine, or spinal fluid:

A. Detection

The automated detection of microbial populations in test samples which are usually sterile, such as blood and spinal fluid, is the simplest function impedance-monitoring instruments can perform. An accelerating change in impedance is known to have an extremely high probability of being associated with a growing population of microorganisms, and thus blood and spinal fluid cultures can be designated positive or negative.

B. Estimating Initial Concentration

This technique can be used in screening nonsterile fluids, such as properly collected, clean-catch urine specimens. The question asked is not how many organisms are present, but rather whether the sample contains more than some defined number of organisms. Samples are then classified as positive or negative on the basis of whether the impedance detection time is less than or greater than a defined detection time corresponding to the concentration of organisms of interest.

C. Antibiotic Susceptibility Tests and Minimum Inhibitory Concentrations

Antibiotics, by reducing growth rates or by killing microorganisms, greatly prolong or prevent detection by impedance. In some cases, the rate of impedance change in the presence of antibiotics is diminished roughly in proportion to the concentration of the antibiotic present. In other cases, the magnitude of the impedance response is diminished. These effects may be used to estimate antibiotic susceptibility.[1,7]

By employing multiple chambers with successive diminishing concentrations of an antibiotic, the lowest concentration which prevents growth or the minimum inhibitory concentration (MIC) can be obtained by impedance monitoring.[7]

D. Identification

Microorganisms may be characterized by their growth rates in various differential media or in the presence of homologous antisera. Although very little work has been done relating impedance measurements to identification, preliminary studies indicate that identification of some organisms of clinical interest may be possible using impedance-based analogs of this concept.[3,8]

IV. INSTRUMENT DESIGN AND OPERATION

The Bactometer® Microbial Monitoring System series of instruments have two major components — an electric measuring circuit (which is housed in an incubator in the Bactometer® 32 Microbial Monitoring System) (Figure 1) and a data display component which is either a strip chart recorder (Bactometer® 8 and 32 Microbial Monitoring Systems) or a computer terminal (Bactometer® 120 Microbial Monitoring System). The component housing the circuitry also contains female connectors which accept the electrode containing sample and reference chambers. All systems contain an oscillator, amplifier, and comparator circuit, digitizing circuit, and address system. A flow diagram of the Bactometer® 32 Microbial Monitoring System is in Figure 2.

In general, all the Bactometer® Microbial Monitoring Systems use two forms of impedance-measuring chambers — modules and bottles or vials.

The module configuration is a disposable cluster of 16 plastic impedance chambers, each containing a pair of stainless steel vertical electrodes emerging from the floor (Figure 3). Chamber volume is about 2.7 mℓ. The electrodes lead from each chamber to form the male edge of a card connector for easy insertion into the instrument connector. The disposable modules are supplied sterile and wrapped in plastic. Closure for the chambers is made by sterile tape which is provided. Inoculum volumes are usually from several microliters to 1 mℓ.

The bottle or vial configuration consists of an overhead pair of stainless steel wires traversing a rubber septum in a bottle or vial enclosure. The wires traverse the septum and dip into the solution contained in the bottle or vial. Sample volumes from 5 mℓ to 1000 mℓ can be handled in this way. The sterile electrode cap assembly is supplied either alone or with a bottle or vial. Bottles or vials are connected to the instrument by means of a bottle connector accessory or in the Bactometer® 120 Microbial Monitoring System by use of a special incubator.

V. METHODS AND PROCEDURES

A. Detection

To detect microorganisms in a clinical specimen of body fluid using any of the Bactometer® Microbial Monitoring Systems, an aliquot of the sample is aseptically added to either a module, bottle, or vial previously filled with culture medium. The appropriate reference chamber, bottle, or vial is also filled with medium, and both sets of

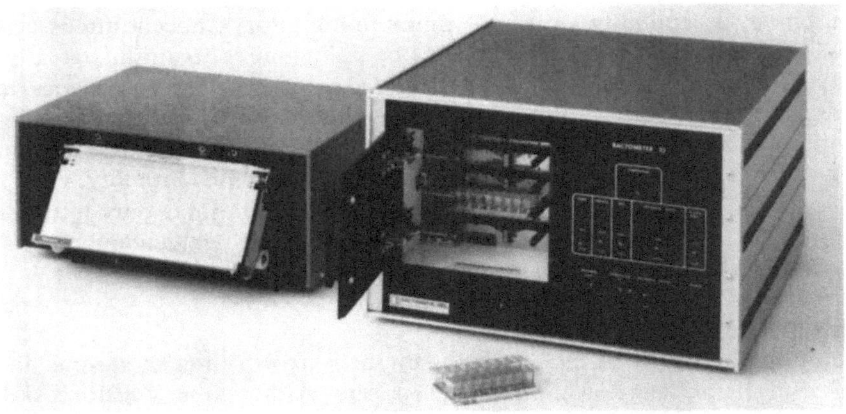

FIGURE 1. Bactometer® 32 Istrument system. (From Bactomatic, Inc. With permission.)

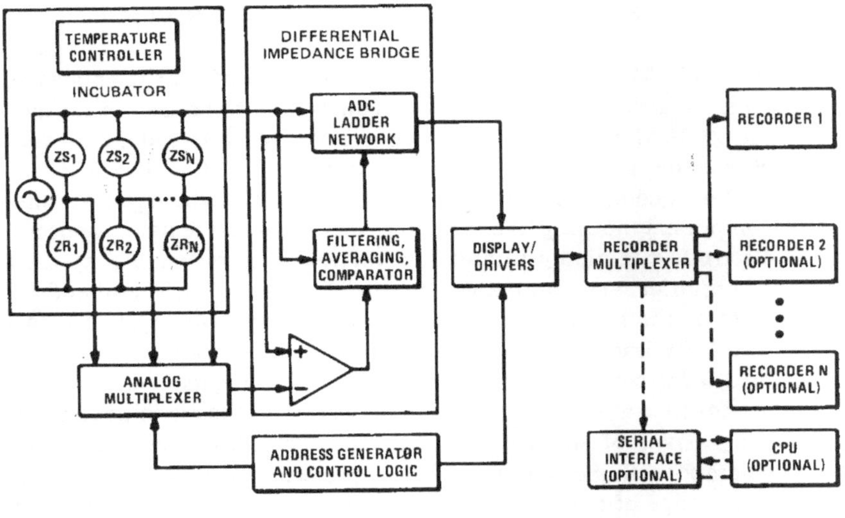

FIGURE 2. Flow diagram of Bactometer® 32 Instrument System. (From Bactomatic, Inc. With permission.)

containers are connected to the instrument which then automatically measures the impedance ratio.

The detection of microorganisms in blood cultures can be accomplished using conventional media in blood culture bottles equipped with impedance electrodes. Blood is collected aseptically as for a routine blood culture and introduced into both an aerobic and anaerobic culture bottle which are then connected to the instrument by use of a special bottle accessory or by use of a special incubator.

To facilitate detection of staphylococci or other organisms which tend to form microcolonies, each bottle is shaken after 1 day. Detection is by strip chart recorder or by microprocessor and is generally sooner than by conventional means.[6,9,10]

Although few clinical trials have been conducted for spinal fluid cultures, it is reasonable to assume that continuously monitoring spinal fluids can speed the detection of the infecting microorganisms. Detection times for the few specimens run to date are approximately 5 to 6 hr.[11]

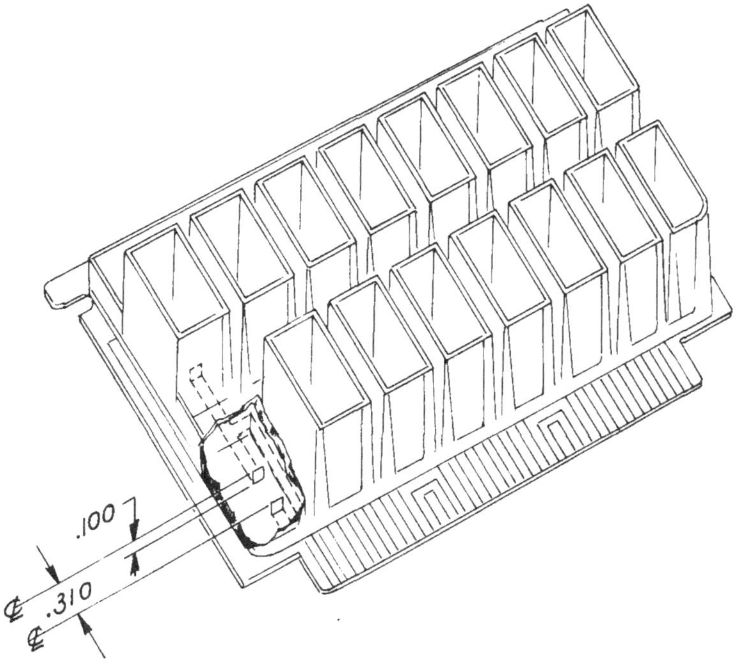

FIGURE 3. Drawing of cuvette for Bactometer® System. (From Cady, P., Dufour, S. W., Shaw, J. and Kraeger, S. J., *J. Clin. Microbiol.*, 7, 266, 1978. With permission. Copyright © 1978, American Society for Microbiology. All rights reserved.)

B. Estimating Initial Concentrations

Impedance measurements can be applied to screening nonsterile samples with high microorganism content for the presence of a specified number of bacteria. For example, one can screen clean-catch urine specimens for those specimens containing greater than 10^5 organisms per milliliter — an indication of bacteriuria.[12] For urine cultures, the procedure is as follows: from a freshly and properly collected urine specimen, 1 mℓ is pipetted into a module prefilled with medium. The module is taped closed and inserted into the instrument and incubated. The detection time is determined by inspection of the strip chart recorder (Bactometer® 8 and 32 Microbial Monitoring Systems) or is automatically displayed by the microprocessor (Bactometer® 120 Microbial Monitoring System). The detection time for common urinary pathogens occurring in a concentration of 10^5 organisms per milliliter or greater is established, and the detection time of the unknown urine specimen is compared to it. Detection times equal to or shorter than the defined time indicate bacteriuria.

C. Antibiotic Susceptibility Tests

A preliminary study indicates that it may be possible to perform antibiotic susceptibility tests using Mueller-Hinton broth containing the appropriate concentration of antibiotic placed into sample and reference chambers. Control blanks are included. An inoculum of organisms is added to the sample chamber which is then connected to the instrument and incubated. Susceptible organisms show prolonged detection times or reduced impedance response in the presence of antibiotics compared to the detection time or impedance response obtained in the absence of antibiotics.

MIC can be determined by dilution techniques in which the organism under test is exposed to antibiotic solutions of decreasing concentration, and the electrical impedance is measured as in antibiotic susceptibility testing. The lowest concentration of antibiotic which prevents evidence of growth under controlled conditions is designated the MIC. It appears after investigation that a 6-hr MIC test or antibiotic susceptibility test could be made practical with further developmental work.[7]

D. Identification

Microorganisms can be characterized by their growth in the presence of various inhibitors and antibiotics or stimulators, such as growth factors and vitamins. Preliminary studies indicate that identification of some organisms of clinical interest may be possible using a matrix of differential media and monitoring the electrical impedance in these media.[3] These measurements of the rates of response in these selective media may give a characteristic pattern for each organism, thus identifying it. It may be possible to identify urinary tract pathogens directly from urine with no prior primary isolation, even with mixed flora, providing that the number of different organisms present is small.[13]

VI. INSTRUMENT FEATURES

There are three Bactometer® Microbial Monitoring Systems with various features and accessories to meet the needs of the user.

A. Bactometer® 8 Microbial Monitoring System

Up to eight samples are monitored automatically and sequentially. The impedance ratio of each sample is displayed and recorded every 3.5 sec. An ink pen strip chart recorder is provided. An incubator is not supplied. Variations in impedance ratio of one part per thousand can be detected.

B. Bactometer® 32 Microbial Monitoring System (Figure 1)

Up to 32 samples are monitored. Display and recording occurs every 3 sec. Data from all 32 samples are displayed and recorded every 96 sec on a heat stylus stip chart recorder. An incubator (30 to 45°C) is provided. Variations in impedance ratio of 1 part per 65,000 can be detected. The instrument can repetitively sample one sample only or one module of eight samples only.

C. Bactometer® 120 Microbial Monitoring System

Up to 480 samples, 120 in each of four modular incubators, are monitored. All data are handled by a microprocessor that interprets the data and then displays the results on a hard-copy printer terminal or on an optional cathode-ray tube video terminal. The incubator is separated into two chambers which can be independently set to any temperature in the range 30 to 45°C. Only one reference chamber is required for 15 sample chambers. This model is intended as the Bactometer® of choice for the clinical laboratory.

VII. INSTRUMENT REQUIREMENTS

Electrical — All Bactometer® Microbial Monitoring Systems require 105 to 125 V AC 60 Hz unless stated otherwise. All models are also available using 208 V AC 50 Hz.

Electrodes — Note that all impedance-measuring electrode characteristics alter with use, and thus reproducible results cannot be guaranteed if electrodes are reused.

Media — All commercial media may be used with the Bactometer® 8 and 32 Microbial Monitoring Systems. Some factory modification of the Bactometer® 120 Microbial Monitoring System may be required to use media with ionic strengths markedly different from normal.

Modules — Modules are guaranteed sterile only if the plastic wrapper is intact.

VIII. MARKET STATUS

The Bactometer® Microbial Monitoring Systems have been on the market since 1976. Urine-screening and blood-culture procedures using the Bactometer® 32 Microbial Monitoring System are available. Antibiotic susceptibility tests are under development while identification is the least developed procedure.

IX. GENERAL COMMENTS

One of the advantages of impedance measurements is the ability to multiplex the use of a single measuring circuit so that a very large number of samples can be measured without physical movement of sample, sensor, or measuring circuit. The Bactometer® 8 Microbial Monitoring System sequentially monitors 8 samples, the Bactometer® 32 Microbial Monitoring System monitors 32 samples, the Bactometer® 120 Microbial Monitoring System monitors up to 480 samples. As microprocessor technology continues to develop, larger systems handling thousands of samples should be feasible.

In three independent studies totaling over 1500 samples, a screening procedure for bacteriuria was used.[12,14,15] Urine samples were analyzed for total plate count and for impedance detection time. Samples containing over 10^5 organisms per milliliter were designated as positive, those with fewer than 10^5 organisms per milliliter as negative. Impedance measurements correctly classified over 95% of the samples and furthermore did so in less than 3 hr. Other papers report that this method has been used successfully for monitoring blood cultures[6,9,10] and blood and spinal fluid.[11] Colvin and Sherris[7] reported that their results with MIC tests were promising; Hardy et al.[16] presented a paper on the use of a modification of the system for anaerobic bacterial analysis; Mischak, et al.[8] reported on the use of impedance measurement for use with detection and identification of mycoplasma.

X. PRACTICAL CONSIDERATIONS

A. Electrodes

1. Geometry of Electrodes

Electrode geometry influences impedance measurements, with larger responses being obtained when the electrodes are closer together. Most variations in electrode geometry have little effect on the measurement. What is important is that the microorganisms be included between the electrodes. This is sometimes not the case when organisms form microcolonies away from closely spaced bottle electrodes or when surface growers are forming on modules filled so that the surface is too far above the electrodes.

2. Electrode Material

Electrodes which give the best response to growing microorganisms are stainless steel and gold. Other metals vary in suitability. All impedance-measuring electrode characteristics vary with use, thus reproducible results cannot be guaranteed if electrodes are reused.

B. Impedance-Measuring Chambers (Modules)

Materials — The material for module fabrication must be nontoxic. Ability to withstand either autoclave temperatures or radiation sufficient for sterilization is preferable.

Configuration — Chambers may be fabricated in a variety of sizes and shapes for microbiological convenience, the only requirement being a septum through which the electrodes can pass.

Capacity — The sample capacity of the various chambers may vary from as little as 1 mℓ to as much as 1 ℓ.

C. Media

Most commercial media which support good microbial growth give satisfactory results with impedance measurements.

D. Strip Chart Display

The Bactometer® 32 Microbial Monitoring System displays 32 samples simultaneously on a single chart. This is accomplished by dividing the chart into 32 bands and using a folded-curve technique.[6] This allows easy visual scanning of all 32 curves for evidence of rapid impedance change.

E. Data Processing

The Bactometer® 120 handles and interprets all data displaying results in a hard-copy printer or cathode-ray tube.

XI. OTHER INSTRUMENT SYSTEMS USING ELECTRICAL IMPEDANCE

A. University of Texas (Goldschmidt and Wheeler)

Another instrument system involving electrical-impedance measurements has been developed by a group at the University of Texas for detecting a bacteriuria directly from the urine without culturing.[17] This is not on the market. This instrument differs from the system described in the previous section in that it has a tetrapolar electrode system and measures the electrical effects of washed bacterial cells suspended in distilled water rather than their metabolites.

Only one reading is made reflecting voltage changes which correlates directly with the number of cells in the test specimen. This method offers rapid results by bypassing growth, colony formation, or the production of metabolites. The instrument system includes a square wave form generator, a volt meter, a 10 kHz band pass, an oscilloscope, four-pronged probe, and steel or platinum electrodes fastened to the cover of a plastic petri dish. In constructing the instrument, the electrical connections are very important. Using this system, there is a correlation between the data and the output voltage in known and unknown specimens. Over 60 clinical urine specimens were prefiltered through 8-μm filters to remove interfering debris, and the bacteria were then washed as usual on the 0.45-μm membrane filters and resuspended in distilled water. Nomographs are formed from the data using known concentrations of organisms. Electrical measurements were made, and the concentrations were calculated by comparing the output voltages with the nomograph. Standard plate counts were used for comparison. The results of the two test methods compared favorably. Impedance measurements on a few mixed cultures could not differentiate them from pure cultures, which may be a limitation of the technique.

This method is most useful as a quick, preliminary screening method to indicate the possibility of a bacteriuria.

B. Harrow, Middlesex, England (Ur and Brown)

Ur and Brown have described an impedance bridge using capillary tube electrodes for sample and reference. Sample size is small — about 100 $\mu\ell$. Electrodes are of gold fabricated by forming an anular ring of gold at each end of the capillary tube. Electrodes are carefully matched in pairs. Early instruments were single channel and not self-balancing. A self-balancing, six-channel instrument will soon be available. Threshold levels are reported to be as low as 10^4 organisms per milliliter.[4,18,19]

Much of the information in this chapter is summarized briefly in Tables 1 to 3.

REFERENCES

1. **Cady, P.**, Progress in impedance measurements in microbiology, in *Mechanizing Microbiology*, Sharpe, A. N. and Clarke, P. S., Eds., Charles C Thomas, Springfield, Ill., 1978, chap. 14.
2. **Cady, P., Dufour, S. W., Shaw, J., and Kraeger, S. J.**, Electrical impedance measurements: rapid method for detecting and monitoring microorganisms, *J. Clin. Microbiol.*, 7, 265, 1978.
3. **Cady, P.**, Rapid automated bacterial identification by impedance measurement, in *New Approaches to the Identification of Microorganisms*, Hedén, C. G. and Illéni, T., Eds., John Wiley & Sons, New York, 1975, 73.
4. **Ur, A. and Brown, D. F. J.**, Monitoring of bacterial activity by impedance measurements, in *New Approaches to the Identification of Microorganisms*, Hedén, C. G. and Illéni, T., Eds., John Wiley & Sons, New York, 1975, 61.
5. **Goldschmidt, M. C. and Wheeler, T. G.**, Determination of bacterial cell concentration by electrical impedance measurements, presented at 73rd Annu. Meet. Am. Soc. Microbiology, Miami Beach, May 1973, 9.
6. **Hadley, W. K. and Senyk, G.**, Early detection of microbial metabolism and growth by measurement of electrical impedance in *Microbiology-1975*, Schlessinger, D., Ed., American Society for Microbiology, Washington, D.C., 1975, 12.
7. **Colvin, H. J. and Sherris, J. C.**, Electrical impedance measurements in the reading and monitoring of broth dilution susceptibility tests, *Antimicrob. Agents Chemother.*, 12, 61, 1977.
8. **Mischak, R. P., Cady, P., and Kraeger, S. J.**, Detection of mycoplasma by impedance measurements, presented at the 75th Annu. Meet. Am. Soc. Microbiology, New York, April 1975, 207.
9. **Specter, S., Throm, R., Strauss, R., and Friedman, H.**, Rapid detection of bacterial growth in blood samples by a continuous monitoring electrical impedance apparatus, *J. Clin. Microbiol.*, 6, 489, 1977.
10. **Kagan, R. L., Schuette, W. H., Zierdt, C. H., and MacLowry, J. D.**, Rapid automated diagnosis of bacteremia by impedance detection, *J. Clin. Microbiol.*, 5, 51, 1977.
11. **Khan, W., Friedman, G., Rodriguez, W., Controni, G., and Ross, S.**, Rapid detection of bacteria in blood and spinal fluids in children by electrical impedance method, presented at the 76th Annu. Meet. Am. Soc. Microbiology., Atlantic City, May 1976, 37.
12. **Cady, P., Dufour, S. W., Lawless, P., Nunke, B., and Kraeger, S. J.**, Impedimetric screening for bacteriuria, *J. Clin. Microbiol.*, 1, 273, 1978.
13. **Cady, P.**, personal communication.
14. **Zafari, Y. and Martin, W. J.**, Comparison of the Bactometer Microbial Monitoring System with conventional methods for detection of microorganisms in urine specimens, *J. Clin. Microbiol.*, 5, 545, 1977.
15. **Throm, R., Specter, S., Strauss, R., and Friedman, H.**, Detection of bacteriuria by automated electrical impedance monitoring in a clinical microbiology laboratory, *J. Clin. Microbiol.*, 6, 271, 1977.
16. **Hardy, D. L., Cox, M. E., and Lindberg, L. H.**, Automated monitoring of anaerobic bacteria by impedance measurements, presented at the 76th Annu. Meet. Am. Soc. Microbiology, Atlantic City, May 1976, 42.
17. **Wheeler, T. G. and Goldschmidt, M. C.**, Determination of bacterial cell concentrations by electrical measurements, *J. Clin. Microbiol.*, 1, 25, 1975.
18. **Ur, A. and Brown, D. F. J.**, Rapid detection of bacterial activity using impedance measurements, *BioMed. Eng.*, 9, 18, 1974.
19. **Ur, A. and Brown, D. F. J.**, Impedance monitoring of bacterial activity, *J. Med. Microbiol.*, 8, 19, 1975.

TABLE 1

Instrument Designation, Principle of Action and Applications of Bactometer® Systems

Instrument trade name	Sponsoring company	Principle of action	Principle of action applied to: (XX = principal application, X = secondary or potential use)				
			Detection	Enumeration	Antibiotic susceptibility	MIC[a]	Identification
Bactometer® 8 Microbial Monitoring System	Bactomatic, Inc.	Impedance ratio measurement	X Onset of impedance change	X Time required for detection of impedance change	XX Relative onset of impedance	X Same principle as antibiotic susceptibility using various concentrations of antibiotics	X Differential response across multiple media
Bactometer® 32 Microbial Monitoring System	Bactomatic, Inc.	Impedance ratio measurement	Onset of impedance change	Time required for detection of impedance change	Relative onset of impedance	Same principle as antibiotic susceptibility using various concentrations of antibiotics	Differential response across multiple media
Bactometer® 120 Microbial Monitoring System	Bactomatic, Inc.	Impedance ratio measurement	Onset of impedance change	Time required for detection of impedance change	Relative onset of impedance	Same principle as antibiotic susceptibility using various concentrations of antibiotics	Differential response across multiple media

[a] Minimum inhibitory concentration.

TABLE 2

Requirements and Features of Bactometer® Systems

	Requirements			Features					
Instrument trade name	Pure culture	Special media	Other	Degree of automation	Test period	Number of tests per day per unit	Accuracy compared to conventional tests	Strengths	Weaknesses
Bactometer® 8 Microbial Monitoring System	Not for detection or enumeration; needed for identification or susceptibility tests	No	Disposable modules or bottles/vials with disposable electrodes	Measurements automated; strip chart interpretation required		Urine, 1—16	Not done[a]	Strengths of all systems: rapid results, ease of use and labor savings; eliminates testing of negative samples	Weakness of Bactometer® 32 strip chart interpretation; pure cultures required for susceptibility and identification
Bactometer® 32 Microbial Monitoring System	Not for detection or enumeration, needed for identification or susceptibility tests	No	Disposable modules or bottles/vials with disposable electrodes	Measurements automated; strip chart interpretation required	Urine 3 hr Blood, 3—7 days	Urine 1—64 Blood, 16/week	Good[a] Good[b]		
Bactometer® 120 Microbial Monitoring System									

[a] All clinical data has been done on the Bactometer® 32 Microbial Monitoring System. Comparable data is expected for the other instruments. Blood cultures are not recommended for the Bactometer® 8 Microbial System.

[b] Data obtained with a one-bottle system. Comparable or better data expected for a two-bottle system.

TABLE 3

Market Aspects of Bactometer® Systems

Instrument trade name	Clinical trials	FDA status	Instrument price: Sale ($)	Instrument price: Lease/rent	Consumables cost: Per vial	Consumables cost: Per test	Commercial potential
Bactometer® 8 Microbial Monitoring System	No	Unclassified	2380	Not available	38c	75c[a]	The Bactometer® Microbial Monitoring Systems are general-purpose instruments and have potential for clinical, food, dairy, water, environmental, and industral microbiological applications
Bactometer® 32 Microbial Monitoring System	Urine and blood cultures	On the market	12,500	Lease available	38c 65c Electrode only	75c[a] $3.00[b]	
Bactometer® 120 Microbial Monitoring System	No	Unclassified	20,000—65,000	Lease available	65c Electrode only 20c	$3.00[b] 40c[a]	

[a] For duplicate urine samples.
[b] For two-bottle test, including electrodes and media.

Chapter 12

EVALUATION OF INSTRUMENTED SYSTEMS TO MEET GOALS FOR MICROBIOLOGICAL ANALYSIS OF BODY FLUIDS

I. THE GOALS

The overall goal of the instrumented systems for microbiological analysis of body fluids is to perform a complete microbiological analysis on all types of body fluids rapidly, accurately, and at a reasonable cost.

A. Ideal Instrument

The ideal instrumented system for microbiological analysis of body fluids will have the following characteristics:

1. Perform all three phases of a microbiological analysis, i.e., detection, quantitation, antibiotic susceptibility and identification on all types of body fluids.
2. Perform the tests as, or more, accurately than traditional methods and allow flexibility for new methodology resulting from state-of-the-art advances.
3. Perform the tests rapidly without the prolonged incubation or need to isolate the predominating microflora.
4. Perform the tests and interpretation with as little manual intervention as possible
5. Present results in a clear, easily understood form.
6. Cost of tests reasonable when compared to other competitive and traditional tests.

The adherence of the instrumental system to the theoretical rules discussed in Chapter 2 and the consideration of the practical aspects discussed in Chapter 3 will aid in the evaluation of the system to meet these goals, both with respect to present capabilities and their future potential.

II. EVALUATION OF THEORETICAL ASPECTS OF MICROBIOLOGICAL ANALYSIS INSTRUMENT SYSTEMS

A. Introduction

Originally, microbiological analytical instruments were designed to measure one parameter and were limited in ranges and restrictions on kinds and conditions of restrictions placed upon the way a sample could be made ready for the instrument. Instruments of this type have their place and are still being developed. However, instruments which combine complete preparation, manipulation, and disposal of the sample, as well as collection, reduction, and interpretation of the data obtained, are in demand. An apparatus which combines all phases of the preparation, processing, and interpretation of pieces of information to obtain precise and accurate knowledge about a phenomenon or condition usually combines several simple instruments and is called a system.

B. Development of Automated Microbiology Analysis System — A Summary

If the goal is to develop a fully automatic system to obtain knowledge which is now obtained from a combination of medical, clinical, and biological laboratories, but in less time with more accuracy and precision, we must first develop a plan of approach for the project. The most important phase in the development of a system is the listing of requirements and restrictions which will be encountered. These should be detailed. Goals which will be necessary to achieve should be clarified and fully understood. The full understanding of a goal requires an in-depth knowledge of the capabilities and actions and the operating ranges and interactions of the phenomena being studied. These ranges may be limited by natural laws or by government legal restrictions. They may be economic, they may be ethical and moral, or they may be technologically beyond the state-of-the-art. Ideally, all of the ranges and limits must be known and understood. Systems have been developed using a method of altering the design parameters as problems are encountered; knowledge is obtained regarding the phenomena being studied and extra capability for the system is required to remain ahead of competing development programs. Systems developed in this way are extremely costly in time and money. A good comprehensive plan which states the problems and reviews the solution options needed to reach clearly defined fixed goals is most important and cannot be overstressed.

The knowledge of the theoretical and practical aspects of the phenomena being observed is reviewed in this chapter and should serve as an example and guide to the kind of thinking which must be part of such a goals-oriented system design and development program. In this chapter, the goals for the design and development of a fully automated, all-use clinical microbiological system are reviewed and tabular summaries are presented of how the aspects of design and rules predicting the natural populations of microorganisms have been utilized in the semiautomated systems discussed in earlier chapters. In Chapter 2, theoretical aspects related to the design and development of systems capable of measuring microbial activity were listed. These aspects can be reflected against the six systems which are presently on or near the market. The capabilities of the PHA system are included for illustrative purposes. The extent to which each system takes advantage of the aspects can be summarized in tabular format by putting a plus for each requirement which is taken advantage of and a minus if the requirement is disregarded, while the block is left blank if the feature is not of direct consequence to the present use of the equipment. The foundation for complete systems must be based upon some set of rules which allows the designer to predict the actions of a natural model under clearly defined conditions. The rules for utilizing the model which consists of microorganisms growing in a fluid medium are therefore restated and reflected against the six systems.

C. Summary Tables

1. Table 1. Properties of Microorganisms Growing in Fluid Medium

Rules which define the properties of live microorganisms in a fluid medium and make possible the prediction of what can be expected of bacteria under stated conditions.

2. Table 2. Comparison of How Well Present Systems Follow Rules in Table 1

Comparision of how well clinical microbiology semiautomated systems on or near the market follow the rules given in Table 1.

3. Table 3. Comparison of How Well Present Instrument Systems Follow Design Considerations

Comparison of how well design considerations involving characteristics of microorganism populations have been used or not by the semiautomatic systems on or near the market.

TABLE 1

Rules for Predicting Reactions of Live Microorganism Populations in a Fluid Medium

1. Concentration Rule — When the minimal cell concentration needed for cell reproduction is equalled or exceeded, the initial lag interval following the inoculation from a stock bacterial population is independent of the number of organisms in the inoculum.
2. Dilution Rule — A population of bacteria growing in log phase may be diluted with medium having exactly the same chemical, physical, and colligative properties as the initial growth medium without interrupting or altering the log growth rate of the population.
3. Equilibrium Rule — Any population of bacteria from a fluid source is in a potential growth equilibrium state that is controlled by all other biological and chemical constituents contained along with it.
4. Reaction Rule — A bacterium in log phase growth under fixed conditions will respond to a controlled applied stress or insult in a predictable manner.
5. Dimension Rule — All microorganisms from a common source cell and growing under the same conditions have exactly similar physical dimensions and metabolic activity at the instant following cellular division.

TABLE 2

Comparison of How Well Present Systems Follow Rules in Table 1

	Concentration	Dilution	Equilibrium	Reaction	Dimension
Bactec®	+		+	+	
Bactometer®	+		+	+	
Autobac 1®		−	+	−	−
Differential III®		−	+	+	
AMS®	+		+	+	
MS-2®		+	+	+	
PHA	+	+	+	+	+

Note: +, system takes advantage of rule; −, system violates rule; if left blank, the characteristic is not of direct consequence to the present use of the equipment.

TABLE 3

Comparison of How Well Present Instrument Systems Follow Design Considerations

	Minute-by-minute, cell-by-cell	Sensing principle unequivocal	Exploits potential sensitivity	All interpretations in log phase	Direct inoculation from body fluid
Bactec®		+	+	+	+
Bactometer®			+	+	+
Autobac 1®		−	+	−	
Differential III®	+		+	+	
AMS®		−	+	+	+
MS-2®			+	+	
PHA	+	+	+	+	+

Note: +, advantage is taken of the population characteristic; − the characteristic is not satisfied and would be desirable in their application; if left blank, the characteristic is not of direct consequence to the present use of the equipment.

4. Table 4. Comparison of How Well Present Instrument Systems Take Advantage of Logic and Circuit Considerations

Comparison of how well the systems on or near market were designed to take advantage of circuit and logic considerations.

In Tables 2, 3, and 4, a plus (+) is given if advantage is taken of the category. The block is left blank if the utilization of the category is not of direct consequence to the present use of the equipment, and a minus (−) is given if the characteristic is not satisfied and would be desirable in the application.

D. Benefits of a Complete System

The extra benefit obtained from a complete system approach is that new measurements are possible. If the system is capable of real time analysis, new applications are technically possible with the addition of dedicated computers. Examples of applications to clinical and industrial microbiology programs which are possible and easy for a real time analysis system were given in the final part of Chapter 2. In our evaluation of each of the six microbiology systems on or near the market, it was noted whether or not the approach that was being used could be adapted in some way to provide real time data on a minute-by-minute basis and on a cell-by-cell basis. Of the six, Differential III® alone has that possibility and this fact is noted in Table 3. Therefore, the system concepts which are part of the Differential III® development make it possible for Science Spectrum to develop systems which can be applied to new applications in clinical microbiology.

III. DESIRABLE FEATURES FOR AND EVALUATION OF INSTRUMENTED SYSTEMS — PRACTICAL ASPECTS

The most desirable features discussed under each heading in the chapter on practical aspects will be summarized briefly, and each of the instrumented systems will be evaluated on these aspects in the following sections. PHA will be commented on where appropriate at its stage of development.

A. Biological

1. Test Culture Features

a. Sample preparation should be simple, designed to offer a reasonable inoculum concentration and to remove interfering substances without introducing new inhibitory substances or adverse environmental factors.
b. The inoculum should be tested to insure a uniform, reasonable inoculum concentration of a culture in uniform logarithmic growth. The approach to be ideal for a rapid method should be able to analyze cultures obtained from the original specimen without preliminary purification, i.e., to analyze mixed cultures without the loss of time needed to perform a primary isolation. If clumping or chaining of bacterial inoculum will influence the results, the method should include a means of breaking up or preventing formation of these aggregates.
c. The source of the inoculum is preferably from the specimen, but with most approaches, replicates of single, fresh "like" colonies from a primary isolation plate may be satisfactory for antibiotic susceptibility tests and identification.

2. Evaluation of Instrumented Systems to Meet Test Culture Features

a. Based on the summary in Table 5, an evaluation may be made of how well each

TABLE 4

Comparison of How Well Present Instrument Systems Take Advantage of Logic and Circuit Considerations

	Sensor self-test	High selectivity for mixed population	Possible 3° of freedom	Near real time change of growing population	1-min process time	Decontamination and test for decontamination	Interpreted output from system
Bactec®	+	−	−	−	+	+	+
Bactometer®	+	−	−	+	+	+	+
Autobac 1®		−	−	−	+	+	+
Differential III®	+	−	+	−	+	+	+
AMS®		+	+	−	+	+	+
MS-2®		−	−	+	+	+	+
PHA	+	+	+	+	+	+	+

Note: +, system offers feature; −, system does not provide feature; if left blank, the characteristic is not of direct consequence to the present use of the equipment.

TABLE 5

Practical Aspects Summary; Biological Considerations; Test Cultures

	Inoculum					
Instruments	Prep (simple)	Size (controlled)[b]	Log growth[b]	Mixed culture	Clump affect	Source (specimen)[a]
Detection[a]						
Bactec® 460	Yes	NA	NA	No	No	Original
AMS® (ID also)	Yes	NA	NA	Yes	No	Original
Bactometer®	Yes	NA	NA	No	No	Original
PHA	Yes	NA	NA	Yes	Yes	Original
Susceptibility[a]						
Autobac 1®	Yes	Yes	Nonlog	No	Yes	Several colonies from primary isolation plate
Differential III®	Yes	Yes	Log	No	No	Several colonies from primary isolation plate
MS-2®	Yes	Yes	Log	No	No	Several colonies from primary isolation plate
AMS®	Yes	Yes	Late log or stationary	Yes	No	From ID well
PHA (ID also)	Yes	Yes	Log	Yes	Yes	From detection vial

[a] ID, identification.
[b] NA, not applicable.

instrumented system meets the desirable qualities for test cultures. All of the instrumented systems that are used principally for detection start with an inoculum that is a simple preparation of the original specimen, probably with the bacteria in the lag phase and with an unknown concentration of microorganisms, both of which conditions are unavoidable in detection procedures from original specimens and must be allowed for in the test methods. The detection systems have mainly demonstrated their capability in the clinical laboratory with blood (Bactec®) and urine (Bactometer®, AMS®, and PHA).

b. Of the instrumented systems performing antimicrobial susceptibility tests or minimum inhibitory concentration (MIC), Differential III® and MS-2® use inocula that are brought to the log phase before exposure to the antibiotics, but Autobac 1® uses the colonies directly from the plate in stationary phase for inoculation of the antibiotic containing cuvettes. This may contribute to nonagreement with Kirby-Bauer of certain organism/antibiotic combinations as noted in their package circular. AMS® derives the inoculum from the positive identification culture, which is probably not in the rapid log phase when first exposed to the antibiotic. Although Autobac 1®, AMS®, and MS-2® test the inoculum for concentration of bacteria before adding antibiotics, only MS-2® of these three retests the inoculum after it has reached log phase to offer positive control of the concentration of bacteria immediately before the antibiotics are added, which is a desirable feature. The PHA method also checks and adjusts the concentration of the culture in rapid log growth before adding the antibiotics or antisera.

c. Two methods of detection have been shown to be feasible to recognize mixed cultures, as found in clean-catch urines, and to perform susceptibility and identification tests without an initial purification — the PHA by virtue of formation of individual peaks for each of the bacteria present and the AMS® by separating the bacteria in the original mixture using selective inhibitory media.

This is an important plus value for these systems, particularly as the other two instrumented systems used primarily for detection, Bactec® and Bactometer®, cannot distinguish mixed cultures, and Autobac 1®, Differential III®, and MS-2® require pure cultures for antibiotic susceptibility and MIC tests. This becomes especially important as these three systems are inoculated with several "like" colonies picked from an overnight primary isolation plate, which may yield a mixed inoculum if the "like" colonies are different, requiring an addition of a streak plate to the procedure to confirm purity delaying the final report for 24 hr. The AMS® derives its pure inoculum from the positive identification media. To keep it pure, the test is performed in that same inhibitory media, which means testing an inhibitory (antibiotic) reaction in the presence of an inhibitor not used in treatment, a seemingly undesirable practice. The PHA susceptibility, as well as identification tests, can be performed directly on the culture obtained from bacteriuria detection, as the peak formed by each bacterium will react individually to the added reagent (antibiotic or antiserum). It should be pointed out that the recognition of a mixed culture can be made in each of the three steps of the analysis with PHA, offering a cross-check on the results.

d. Clumping or chaining of the bacteria in the inoculum can cause errors semiquantitation and serological agglutination methods using the PHA unless these aggregations are prevented or separated as by agitation. Also, product literature advises in reading Autobac 1® results that the cuvette should be inspected for clumping by certain types of cultures to make sure the growth has not settled out.

The aggregation of organisms tends to give low plate counts in the conventional counting procedures. This tends to give the illusion of "false positives" for instrumented methods, especially those which depend on metabolic end products when compared to plate counts.

3. Desirable Features for Test Methods

The test methods will vary widely with the chosen approach, but there are several common aspects desirable in all methods.

a. The test environment is very important, and the ideal instrumented system will employ test methods that will combine a suitable incubator with agitation, the appropriate gaseous atmosphere, and the best media formulation for rapid growth compatible with all test reagents to be used.
b. The most desirable test methods are those designed to make measurements that take full advantage of the potential of the chosen approach to perform rapidly a complete microbiological analysis, including detection/enumeration, antibiotic susceptibility tests and MIC, and identification.
c. The flexibility of the instrumented methods to perform all types of analyses on all types of body fluids is valuable.
d. Adequate controls by the test method on the analytical and decision-making processes are vital.

4. Evaluation of Instrumented Systems re Test Methods

a. The adequacy of implementation of the test method for each approach shows some variation with regard to the test environmental features that are an integral part of the instrumented system. (See Table 6). All but two systems supply an incubator (Differential III® and Bactec® 460 do not, Bactec® 225 does), but only Autobac 1® and MS-2® (and PHA projected) supply agitation. Proprietary media are required for all but two systems, Bactometer® and Differential III®. An environmental gas system is available with the Bactec® 460, providing a suitable atmosphere for anaerobic bacteria and microorganisms responding to CO_2.

b. All of the test methods seem to employ the capability of the various approaches well. Some instrument systems originally were aimed primarily at detection of an infection in body fluids, Bactec® for blood and Bactometer® for urines, while others are designed principally for antibiotic susceptibility and MIC tests on pure cultures (Autobac 1®, MS-2®, and Differential III®). AMS® and PHA have been shown to be capable of across-the-board determinations, mainly on urine specimens at this time.

c. Almost all of the instrumented systems have the potential to be modified to perform a complete analysis with pure cultures. These modifications are being made and feasibility studies conducted with the view to capturing a market broader than the original claims. However, since one of the more important goals of a rapid method of microbiological analysis is the elimination of the purification step with its overnight incubation, the two instrument systems, AMS® and PHA, which have the potential to handle mixed cultures, have an advantage. One further consideration is the capability of the various instrumented systems to handle body-fluid specimens containing a multiplicity of normal microflora as well as possible pathogens. To date, blood, urine, and spinal fluids have been the main fluids analyzed. These specimens are normally sterile or minimally contaminated, and mixed cultures of two bacteria are manageable in AMS® and PHA. The prospects of even these instruments to handle three or more bacteria in large numbers is yet to be established.

d. Some approaches do not seem to lend themselves to expansion to susceptibility testing and identification without the purification process. These include Bactec®, Bactometer®, Autobac 1®, Differential III®, and MS-2®. The former two depend entirely on metabolic responses, and the latter three would require a pattern of metabolic activity using selective media or a growth/no growth profile using selective inhibitory media to establish identification. It is not practical to do identification when two separate types of bacteria are making metabolic inputs in the same culture. Unless selectively inhibitory media are used throughout, the results would be very difficult to interpret to establish an identification.

5. Desirable Features in Data Measurement

a. The sensitivity of the data measurements and the signal-to-noise ratio must be sufficiently large to yield clear-cut results.

b. The data measurement intervals are most effective when geared to the major dynamic changes in the reaction.

c. Several clear-cut criteria of a reaction are required for an adequate analysis and may be enhanced by manipulation of the raw data.

d. Reports with a traditional phraseology and in simple form are most easily applied to diagnosis and treatment by the physician.

e. Interpretation of the data must include an assessment of the significance of a false negative vs. false positive results as well as magnitude of error.

TABLE 6

Practical Aspects-Summary Biological Considerations of Test Methods

		Environment (provided)				Analytical functions (Principal XX)				
Instruments	Approach	Incubator	Shaker	Media	Gas	Detection	Enumeration	Antibiotic susceptibility	MIC	Identification
Detection										
Bactec® 460	Radiometric	No	No	Yes	Yes	XX	X	X[a]	(X)	X[b]
AMS® (ID also)[c]	Light transmission	Yes	No	Yes	No	XX	XX	X	X	XX[d]
Bactomer®	Impedance	Yes	No	No	No	XX	XX	X	X	X[e]
Susceptibility										
Autobac 1®	Light absorbance	Yes	Yes	Yes	No	X	X	XX	XX	X[e]
Differential III®	Differential light scatter	No	No	No	No	X	X	XX	XX	X
MS-2®	Turbidity	Yes	Yes	Yes	No	X	X	XX	XX	X[e]
AMS®	Light transmission	Yes	No	Yes	No	XX	XX	XX[f]	X	XX[d]
PHA (all analyses)	Pulse height analysis	Yes	Yes	Yes	Yes	XX	XX	XX	(X)	XX
		(Projected)								

Note: XX indicates principal analytical function; X indicates secondary function; (X) indicates potential.

[a] Serum assay (aminoglycosides).
[b] Selective substrate.
[c] ID, identification.
[d] Inhibitory media.
[e] Differential media.
[f] *Escherichia coli*

6. Evaluation of Systems re Data Measurement

The adequacy of data measurement is tied not only to the instrumented approach used, but to the adherence to biological aspects of the testing. There are several factors that affect the quality of the data measurement which are summarized in Table 7.

The sensitivity of the data measurement may be judged by the excess signal-over-"noise" ratio, so that the results are not marginal. Most of the instrumented systems used to detect bacteria will detect 10^5 bacteria per milliliter, which seems to be an adequate sensitivity for meaningful interpretation.

The nature of bacterial growth makes it ideal for analysis by a dynamic process. All of the detection methods require several readings which produce several progressive data points for analysis. Of the instrumented systems determining antibiotic susceptibilities or MIC tests, Autobac 1® takes only one reading, while MS-2®, AMS®, and PHA take more than one reading, as does Differential III®.

The reading intervals also vary with the approach used and type of sample. Some, such as Bactometer®, read continuously; MS-2® reads at 5-min intervals, and AMS® reads 4 times per hour, reading each specimen once an hour. As long as enough readings are taken to yield several definitive data points which span the period of greatest reactivity or change in data, the exact frequency of data measurements may not be too critical.

The total test period varies with the approach and the nature of the specimen being tested. Bacteriuria can be detected in 2½ to 3 hr by Bactec®, Bactometer®, and PHA but with AMS®, the specimen is not deemed negative for 13 hr. Detection of a bacteremia by Bactec® requires several hours to several days, depending upon the type of microorganism and the initial concentration of bacteria in the blood specimen. Autobac 1® and MS-2® determine antibiotic susceptibilities in 3 to 5 hr, while Differential III® has results in 90 min and PHA in 1 hr. AMS® requires 12 hr for the same test.

The number of criteria desirable to yield definitive data has been discussed thoroughly elsewhere, but it should be emphasized that at least three criteria are considered to be required for optimum results. At present, only the PHA method produces at least three criteria for interpreting each reaction.

B. Engineering

1. Desirable Features, Functions and Automation, and Design

a. The most desirable instrumented systems will be able to perform all of the functions required for one phase of analysis or ideally the complete microbiological analysis. The number of processes performed by the instrumented system determines the degree of automation of the system.
b. The most important functions to automate are the measurement of the data, the reduction and interpretation of the data, and decision-making based on the data. Automation of other functions, such as incubation/agitation, inoculation, transference of sample, and addition of test substances for continuation of the analysis is highly desirable in a sophisticated instrumented system, while sample preparation and disposal are less necessary.
c. The more easily the instrument design can be modified to meet changing needs, the better.

2. Desirable Design Features

a. The performance characteristics of the instrumented system should be based on

TABLE 7

Practical Aspects-Summary Biological Considerations Data Measurement

Instruments	Detection; sensitivity	Dynamic measurement	Total test period	Frequency of reading	Criteria
Detection					
Bactec®	$10^5 - 10^6$	Yes	Blood, several hours to several days; urine, 1—3 hr	Blood, aerobic 4 hr; repeat 4 hr, twice (AM&PM) day 2; then 3, 4, 5 and 7 days. Anaerobic-days 2, 3, 4, 5, 7 and 10; urine-1—3 hours	Radioactivity increased
AMS® (also ID)[a]	7×10^4	Yes	13 hr (bacteriuria)	1 time per hour	Light transmission reduced
Bactometer®	10^5	Yes	2—6 hr (bacteriuria)	Continuously	Impedance changes
PHA	10^5	Yes	2½ hr (bacteriuria)	3 times-start, 1½ and 2½ hr	1. Total count × 2 2. Channel 255 count 2 3. Emerging pattern above channel 30
Susceptibility					
Autobac 1®	NA[b]	No	3—5 hr	Once	Emerging light reduced
Differential III®	NA	Yes	90 min	Twice-start and end	Differential light scattering intensity
MS-2®	NA	Yes	3—5 hr	5-min intervals	Turbidity
AMS®	NA	Yes	12 hr	1 time per hour	Light transmission reduced comparably to control
PHA (susceptibility) (Criteria 1 and 2)	NA	Yes	1 hr	Twice-start and end	1. Total count lowered 2. Peak count lowered
ID (All four criteria)	NA	Yes	1 hr (30 min—1 hr)	Twice-start and end (20— 30 min)	3. Large particle increased 4. Proportionate pile-up of large particles

[a] ID, identification.
[b] NA, not applicable.

TABLE 8

Practical Aspects-Summary; Engineering Automation of Processes-Detection

	Bactec® 460[a]	Bactec® 225[b]	AMS® (and ID)[c]	Bactometer® 32[a]	Bactometer® 120/480[b]
Sample prep	–[d]	–	M[e]	–	–
Load filler	–	–	M	–	–
Inoculation	M	M	A[f]	M	M
Load incubator (or instrument)	M	M	M	M	M
Incubation	(Separate unit)	A	A	A	A
Loading instrument	M	–	–	–	–
Reading	A	A	A	A	A
Data reduction	A	A	A	A	A
Interpretation	A	A	A	M	A

[a] Most used instrument in manufacturer's series.
[b] Most automated instrument in manufacturers' series.
[c] ID, identification.
[d] –, not necessary.
[e] M, manual.
[f] A, automated.

the most compatible combination of speed, accuracy, reliability, and reproducibility. Since biological reactions are notoriously subject to variability, the recommended judgment on the instrumented vs. the traditional methods should be based on whether or not the instrumented results are as good (or better) than the traditional results rather than on absolute accuracy, etc.

b. The configuration of the instrumented system is dependent on the type of approach used, the complexity of the processing, and the number of specimens that can be accommodated in the cuvettes and the overall system. The instrument system configuration should adequately house the chosen approach to the microbiological analysis and should accommodate (with additional modules, if necessary) a normal day's workload and stay within the weight, power, and size parameters available in the laboratory

c. The operational features and limitations must consider the safety of the operator and protect the environment and culture from contamination by bacteria or other noxious agents. The instrument's ease of operation, maintenance, and serviceability are important, especially to the user.

3. Evaluation of the Instrumented Systems for Functions and Automation and Design

a. All of the instrumented systems perform several functions needed for one phase of the analysis (i.e., detection, susceptibility test, etc.) and two, PHA and AMS®, have shown feasibility to perform the complete analysis.

b. The degree of automation has been summarized in Table 8 for each instrument used primarily for detection and in Table 9 for antibiotic susceptibility or MIC. The detection systems in general use manual operations until the incubation step, after which the data collection and management are automatic. The AMS® performs an instrumented inoculation, and Bactec® 460 requires manual loading of the incubator followed by manual transfer to the instrument for reading. There is more variation in the automation of the susceptibility tests. All instrumented systems depend mainly on manual or semimanual operations to prepare,

TABLE 9

Practical Aspects-Summary; Engineering Automation of Processes-Antibiotic Testing

	Autobac 1®	Differential III®	MS-2®	AMS®
Pure culture	M[a]	M	M	A[b]
Transfer	M	M	M	M
Preincubation	–[c]	M	–	–
Load fill module	–	–	–	M
Fill cuvettes	SM[d]	M	SM	A
Add antibiotics	A[e]	A[e]	A[e]	A
Load into instrument	–	–	M	M
Standardize inoculum	M[f]	–	A	M
Inoculate	M	M	A	A
Load into incubator (or instrument)	M	M	–	–
Incubation	(Separate unit)	(Separate unit)	A	A
Load instrument	M	M	–	–
Reading	A	A	A	A
Data reduction	A	A	A	A
Interpretation	A	A	A	A

[a] M, manual.
[b] A, automated.
[c] –, not necessary.
[d] SM, semimanually — all chambers filled in one operation.
[e] added with separate automated device.
[f] photometer in instrument.

transfer and preincubate the inoculum, and fill the cuvettes, except AMS® which automatically purifies the culture and provides an instrumented system to inoculate the media-filled cuvettes. All systems use an instrumented device to add the antibiotics or provide cuvettes prefilled with antibiotics. Only MS-2® automatically controls the concentration of the test culture in log growth directly before adding it to the antibiotics. After inoculation of the antibiotic containing cuvettes, MS-2® and AMS® finish the analysis automatically, while both Autobac 1® and Differential III® have several manual steps before reaching the automated functions of data measurement and management.

Thus, all the instrumented systems provide automation for the analytical phases of data measurement, processing, and interpretation which benefit the most from automation, while there is considerable variation in the functions less aided by automation. Each individual situation must be evaluated to determine the desirability of the automation offered by the various instrumented systems.

c. The performance characteristics seem to be good for all instrumented systems. However, if test rate is gauged by the time to completion of the various phases of the analysis or the whole analysis, there is considerable variation among instrumented systems as noted in Table 7, where detection of bacteriuria takes approximately 2½ hr for Bactec®, Bactometer®, and PHA, while AMS® requires 13 hr and antibiotic testing takes 60 min for PHA, 90 min for Differential III®, 3 to 5 hr for Autobac 1® and MS-2®, and 12 hr for AMS®.

d. The size of the instrumented systems is quite variable, due to the different treatment and completeness of instrumentation of each approach, i.e., some have all-in-one overall housing and others modular. It is difficult to make accurate comparisons. It would appear that AMS® is the largest (it is very self-contained and

TABLE 10

Practical Aspects-Summary Monetary Considerations[a]

Instruments	Cost per instrument ($)	Cost per consumer test: Per vial ($)	Cost per consumer test: Per test ($)
Bactec® 460 (basic system)	12,900 (15, 150 plus data logger)	1.06—1.25	2.12—2.50
AMS® 120	49,500	3.00	3.00; 3.00 each additional susceptibility test
Bactometer® 32	12,500	0.38 0.65 (electrode only	0.75 (duplicate urine) 3.00 (2-bottle test; electrodes and media)
Autobac 1®	22,500	1.06—1.43	1.06—1.43
Differential III®	NA[b]; rental 0.40 for first MIC; 0.20 each additional	0.10	0.10 times number of drugs tested
MS-2®	34,000	3.50	3.50

[a] Prices as of May 1, 1978.
[b] NA, not applicable

does a complete analysis), while Bactec®, MS-2®, Differential III®, and Bactometer® are quite compact (performing only one phase of the analysis). Autobac 1® is intermediate in size between the two extremes.

e. All of the instruments appear to be designed to be safe, easy to operate, service, and maintain.

C. Monetary Considerations

The overall cost of the design, development, and manufacture of instrument system and consumables are primarily of importance to the seller, while the cost per test compared to the traditional procedures is of greatest importance to the consumer. Although prices will be variable with time after publication, a summary of the current prices (May 1, 1978) is made in Table 10. These figures should be used as a basis for comparison only.

1. Instrument Cost

There is a rather wide range in the present cost of the instruments with the two instruments: Bactec® 460 and Bactometer® 32, used mainly for detection, costing the least; those used for antibiotic testing, at the intermediate price; and the AMS® complete analysis system, costing the most.

2. Cost Per Test

The costs per test vary from $0.75 for bacteriuria detection using Bactometer® to $3.50 for an MIC test done using MS-2®.

Both the cost of the instrument (a one-time cost) and the cost per test (a continuing expense) must be evaluated with each individual situation in mind to decide which is the best value. Cost may be of less importance than other considerations, but is a necessary factor in the evaluation.

IV. SUMMARY STATEMENTS

This summary of the practical aspects relating to each of six sophisticated instrumented systems for microbiological analysis of body fluids which is on or near the market is intended as a guide to understanding and evaluating these systems for your own purposes.

It should be stressed that it is not the intention of this monograph to recommend one instrument or downgrade another in this evaluation. Each person must decide what is needed in his unique laboratory situation and identify the instrumented system (if any) which most nearly meets his requirements. A feature offered by one instrument may not be needed in a certain laboratory and therefore not be an advantage, while a limitation of another instrumented system may be of no consequence in a certain situation and could be disregarded.

Perhaps the most critical evaluation of all is "What have instrumented systems contributed to the microbiological analysis of body fluids?" "What can they be expected to contribute in the future?" "Have they met their goals?"

The instrumented systems already on or approaching the market have done much to increase the speed and accuracy of the tests they perform and have eliminated much of the routine drudgery in the laboratory, freeing microbiologists to concentrate on the difficult, less routine analyses. The speed of the analyses, both with respect to the time spent performing each test and the completion and reporting of the entire analysis, has been stepped up markedly, giving the doctors more prompt diagnostic aid. The accuracy of the analyses has been improved by replacing subjective decisions and interpretations by standardized, nonsubjective measurement of the data. Some approaches have cut down on or appear feasible to eliminate long isolation procedures prior to performing susceptibility and identification tests.

To date, these advantages have not been fully developed. There is no instrumented system that can perform a complete analysis on diverse types of body fluids in a few hours, but many are broadening their scope to do more complete analyses on one body fluid or to analyze a broader spectrum of body fluids.

Perhaps among the real achievements of instrumented systems is that of focusing the attention of the microbiologists on the precise definition of their analyses, resulting in a better understanding of the microbiological processes. The results of instrumented procedures have alerted the microbiologists (and doctors) to the possibility that microbiology can be speeded up, especially by the elimination of the lengthy incubation period, and that microbiology can indeed move into the 20th century.

Chapter 13

LESS SOPHISTICATED INSTRUMENTS

I. INTRODUCTION

There are several instruments or systems on the market that are less sophisticated than those discussed in previous sections. In general, these instruments can perform only parts of a microbiological analysis and cannot make decisions or interpret the data independently. Their chief usefulness is in performing a manual chore automatically. These instruments include microdilutors, which are primarily used for MIC determinations; the Cathra International Repliscan® which mechanizes portions of the traditional methods for identification and antibiotic susceptibility; and gas chromatographs, that are mainly used in the clinical laboratory for identification of anaerobes or bacteria that are difficult to identify by traditional methods.

II. MICRODILUTORS

There are several microdilutors on the market which are in use for MIC determinations. Since the MIC tests are being performed more frequently, this type of instrument may become more common in clinical microbiology laboratories. Two of them will be discussed briefly.

In 1970 MacLowry et al.,[1] working at the National Institutes of Health, reported on a mechanized semiautomated system for microliter measurements of antibiotic sensitivity using the tube dilution method. This device speeded up the manual procedures by a factor of 20 or more.

A. Autotiter 10®

Following this basic work, Canalco, Inc., came on the market and later sold to Ames Company, a Division of Miles Laboratories, Inc., Elkhart, Ind., a series of microdilutors with an increasing degree of automation to perform serial dilution of antibiotics or antisera for MIC determinations or serological tests. Autotiter 10® is the fully automated version which is designed to dilute and dispense the various reagents contained in reservoirs by calibrated diluting loops and dispensing needles which are controlled by a preprogrammed instruction card.

Using the Autotiter® to perform an MIC determination, eight different antimicrobial agents may be tested against two different organisms in one tray or 16 antibiotics against one organism. After diluting and dispensing the antibiotic, the inoculation with the pure culture of the organism must be performed by a separate unit and the test incubated and read manually, as usual.

Two groups of investigators have used the Autotiter® in antibiotic studies and have found the instrument satisfactory for MIC determinations, both with single antibiotics[2] and in combination.[3]

B. Dynatech MIC-2000 System®

A similar automated microdilutor for MIC determinations, tradenamed Dynatech MIC-2000 System,® was introduced on the market more recently by Cooke Laboratory Products, Division of Dynatech Laboratories, Inc., 900 Slaters Lane, Alexandria, Va. This differs from the Autotiter 10® by being more compact and offering an automated bacteriology inoculator as part of the system.

Mitchell et al.,[4] reported on the use of this instrument for identification of Enterobacteriaceae using 20 differential media and performing susceptibility tests on dilutions of 11 antibiotics. In general, the identification and MIC tests were satisfactory.

Therefore, in performing on MIC determination, the role of Autotiter 10® and Dynatech MIC-2000 System® is limited to the elimination of manual dilution and dispensing the test reagents with no decision-making capability.

III. CATHRA REPLISCAN®

The identification of Enterobacteriaceae is one of the major tasks in a clinical microbiology laboratory; and in a previous section, there is a brief discussion of several noninstrumented, rapid methods of identification of Enterobacteriaceae which are on the market, some of them with computer back-up in the laboratory of the parent company. An instrument Repliscan®, a semiautomated processor with its own built-in computer which mechanizes part of the traditional methods for the identification of Enterobacteriaceae and for antibiotic susceptibility testing, was introduced by Cathra International, 1450 O'Connor Drive, Toronto, Ontario, Can., about 3 years ago. This instrument is based on the work of Hinton and Shaeff of the Banting Institute, University of Toronto.[5] The instrument system is composed of the Repliplate®, Replicator®, and Repliscan® or Replireader®.[5]

A. Repliplates®

Repliplate® media are a set of media of two types contained in Petri dishes. The biochemical set of 16 media provides identification of the bacteria by developing the usual color change patterns and sensitivity media, containing precise amounts of antimicrobials (antibiotics) for accurate sensitivity testing, show resistance to the antibiotic by continued culture growth, and susceptibility by lack of growth.

B. Replicator®

Designed to mechanize the laborious procedure of manual inoculation, the Replicator® allows a trained technician to inoculate up to 36 samples of inoculum onto the minimum requirement of 22 Repliplates® in 20 to 25 min. The Replicator® consists of a base with swing arm, which holds an inoculation head of 37 1-mm or 3-mm diameter rods, and a well plate into which the rods are inserted. The organisms are pipetted into the chambers of the well plate and transported to the Repliplate® surface by the inoculation head.

C. Repliscan®/Replireader®

The Repliscan® and Replireader® perform similar functions, providing a method of reading the Repliplate® results and computerizing the response of these results.

Repliscan® is a reading, analytical, and printing unit for obtaining and recording results from the Repliplate®. The plates are loaded into the "Carrier Tray", inserted into the scanner, positioned, and read. Samples are read as positive or negative by the light-sensitive diodes at each window, as interpreted by the operator. This pattern compared electronically to the data bank is recorded, and the sensitivity of the organism is then printed out also. The tray is electronically commanded by the light pen to go to the next position. All readings are completed at the rate of 60 organisms per hour.

For ease of operation and hard copy results, the Replireader® is available with an optional printer that negates the necessity of hand recording the results shown in the light display. The operator places the Repliplates® one at a time on the viewing surface and records the various responses by pushing buttons. The responses at each position are correlated with the data bank, and a numerical identification number appears in a light display.

A group at Mayo Clinic evaluated this system for the identification of gram negative rods in the clinical laboratory. They have purchased the Repliscan® in a modified version.[6] These modifications are comprised of additional patterns to the data bank which the manufacturer has added to all new production units, and additional peripheral equipment in order to increase the MIC capability of the unit.

Repliscan® and its smaller unit, Replireader®, are now available in the U.S. exclusively through Scientific Products, McGraw Park, Ill.

This instrument system carries the identification of Enterobacteriaceae one step beyond the rapid method kits but still requires manual intervention, is valid only for purified cultures, and has no decision-making capability.

IV. Gas Chromatography

One of the earlier methods for identifying bacteria that departs from the traditional procedures is the use of gas-liquid chromatography to separate and graph certain of the various end-products formed by bacteria during growth. Gould[7] showed that the presence of unique end-products or patterns of these substances is often characteristic of a bacterial species and can be used for identification of the bacterium. This determination is usually performed on isolates grown in pure culture but, also, can be done directly on the patient's specimen or serum. Any one of a number of gas chromatograhic instruments can be used, including Capco Anaerobic Identification System Mode 700®, Dohrmann 5-C3®, Beckman GC-2d®, and Packard Instrument Co., Becker 406®.

To identify a bacterial culture, the method in brief consists of growing the culture in a nutrient broth until end-products are formed in measurable quantities and then extracting these metabolites for analysis in the gas chromatograph. The organism is then identified by the pattern of the end-products by comparison with known bacterial metabolic patterns.

To identify the causal organism directly from a specimen or the patient's serum, a test is made for one or two unique end-products formed in the body by the bacterium. This analysis may not require quite as much sample preparation and will eliminate the incubation required for an isolated culture to form end-products, as the end-products will presumably be preformed in the specimen prior to collection.

The determination often requires considerable manual preparation of the sample prior to the development of the chromatographic pattern, and for this reason, cannot be considered as a truly automated rapid technique, even though those tests using the direct estimation of the specimen shortcut the usual incubation period. Gas chromatography has found one of its greatest applications in the identification of anaerobic bacteria with very active development of the technique carried out by Moore and Holdeman at the VPI Anaerobe Laboratory.[8] The method gives accurate results, and it is probably the method of choice for identifying anaerobes in laboratories with a capability for doing gas chromatography.

This technique has been applied to the separation of the genera staphylococci and micrococci by Greenup et al.[9] Direct examination of spinal fluid in meningitis patients has been performed by Controni et al.[10] and Brook et al.,[11] and clinical material has been directly analyzed for streptococci by Mayhew et al.[12] and for anaerobic infection by Maccani.[13] An excellent, brief review of this subject has been written by Mitruka.[14] Recently, Wade et al.[15] have reported on advances in automation of gas-liquid chromatography so that 90% of the characterization procedures of tetramethylammonium hydroxide-treated bacteria is automated.

Certainly, gas-liquid chromatography for bacterial identification is a candidate for complete automation, especially the direct analysis of body fluids, and developments in this field should be carefully considered.

REFERENCES

1. **MacLowry, J. D., Jacqua, M. J., and Selepak, S. T.,** Detailed methodology and implementation of a semi-automated serial dilution microtechnique for antimicrobial susceptibility test, *J. Appl. Microbiol.*, 20, 46, 1970.
2. **Barry, A. L. and Effinger, L. J.,** Evaluation of a semi-automated system for preparing microdilutions for antibiotic susceptibility testing, presented at the 74th Annu. Meet. Am. Soc. Microbiol., Chicago, May 1974, 84.
3. **Ayers, L. W. and Kunke, P.,** A semi-automated method for determining the *in vitro* action of antibiotics in combination, presented at the 7th Annu. Meet. Am. Soc. Microbiol., New York, April-May 1975, 43.
4. **Mitchell, B. L. and Jungkind, D. L.,** Improved identification and susceptibility testing of *Enterobacteriaceae* using a microtiter plate system, presented at the 77th Annu. Meet. Am. Soc. Microbiol., New Orleans, May 1977, 52.
5. KVL Laboratories Sales Manual, Cathra International, Toronto, Canada, August 1977.
6. **Shaw, J.,** personal communication.
7. **Gould, J. R.,** Identification of bacteria by gas chromatography in Proc. Conf. on Rapid Identification of Biological Agents, Davis, I. and Williams, R. E., Eds., April 1966, U.S.A.F. School of Aerospace Medicine, Brooks Air Force Base, Tex., and Office of Naval Research, Washington, D.C., 1967, 25.
8. **Holdeman, L. V. and Moore, W. E. C.,** Anaerobe Laboratory Manual, 3rd ed., Virginia Polytechnic Institute, Blacksburg, Va. 1975.
9. **Greenup, C. J. and Ayers, L. W.,** Application of gas-liquid chromatography to the separation of the genera *Staphylococcus* and *Micrococcus*, presented at the 75th Annu. Meet. Am. Soc. Microbiol., New York, April-May 1975, 30.
10. **Controni, G., Rodriquez, W., Deane, C., Ross, S., Puig, J., and Khan, W.,** Rapid diagnosis of meningitis by gas-liquid chromatographic analysis of cerebrospinal fluid lactic acid, presented at the 7th Annu. Meet. Am. Soc. Microbiol., New York, April-May 1975, 30.
11. **Brook, I., Bricknell, K. S., Overturf, G., and Finegold, S. M.,** Abnormalities in spinal fluid (CSF) detected by gas-liquid chromatography (GLC) in meningitis patients, presented at the 76th Annu. Meet. Am. Soc. Microbiol., Atlantic City, May 1976, 43.
12. **Mayhew, J. W., Onderdonk, A. B., Bartlett, J. G., and Gorbach, S. L.,** Direct gas-liquid chromatographic (GLC) analysis of clinical material for rapid detection of streptococci, presented at the 74th Annu. Meet. Am. Soc. Microbiol., Chicago, May 1974, 74.
13. **Maccani, J. E. and Finn, G. R., Sr.,** Experience using direct gas-liquid chromatography of clinical material to identify anaerobic infection, presented at the 77th Annu. Meet. Am. Soc. Microbiol., New Orleans, May 1977, 59.
14. **Mitruka, B. M.,** Detection and identification of bacteria by gas chromatography in *Methods of Detection and Identification of Bacteria*, Mitruka, B. M. and Bonner, M. J., Eds., CRC Press, Cleveland, 1976, chap. 4.
15. **Wade, T. J., Mandle, R. J., and Schaedler, R. W.,** Further automation of the gas-liquid chromatographic characterization of tetramethylammonium hydroxide-treated bacteria, presented at the 74th Annu. Meet. Am. Soc. Microbiol., Chicago, May 1974, 113.

INDEX

A

B

C

D

E

F

K

L

M

N

O

P

Q

R

S

T

U

V

W